Asociación Civil Eugenio Mendoza
para la Cultura

Teléfono: 571.71.20

THE ORCHIDS
ISBN 980-216-017-2

GALFRID C. K. DUNSTERVILLE

VENEZUELAN ORCHIDS

ARMITANO

FOREWORD

The author of this book, G. C. K. Dunsterville, is a very special person, an example of dedication and a clear sense of purpose, immersed in his world of orchid species. For the past three decades, he has focussed thoughts and action on the search for and study of our native orchid species which, numbering over one thousand, inhabit Venezuela's various ecosystems.

Orchids, as is the case with most plants, only flower at certain times of year and the flower is essential in order to determine the specific place of the plant within the Orchid Family. Nevertheless, Stalky —as he is known to his friends— has not been discouraged. Despite the seasonal obstacle to the study of orchids, he has created something positive and harmonious, binding him closer to his plants and to his marvellous companion, Nora, his wife and colleague of 56 years. Together, they organize and carry out orchid hunts and field work. They carefully observe the ecological niches and special environments where the species live. They record, collect and identify the most common, which they greet with the enthusiasm and joy of meeting old and cherished friends. Those orchids which are unfamiliar or whose identification has not been settled, they bring back, along with the "unknowns" —plants not in flower which are difficult to recognize.

Once collected, the orchids are carefully tended in Nora's conservatory and greenhouses at "Quinta Colibrí", the Dunstervilles' home in Oripoto near El Hatillo, southeast of Caracas. When Nora's care is rewarded by a new flower, then Stalky takes over with his incomparable skill, dissecting the flower and making precise drawings of the plant and its floral organs which allow the species to be classified taxonomically.

It is thanks to this team work and the efforts of other Venezuelan and foreign professionals that our country is proud to call the orchids of Venezuela its best studied plant family, taxonomically speaking, and the richest family in number of native species.

In the pursuit of orchids, Stalky has settled with us permanently in Venezuela, completing 40 years of fruitful work. His international renown as an orchidologist is attested by many scientific honours and, above all, by his books and his numerous articles in Venezuelan and foreign journals.

LEANDRO ARISTEGUIETA

LIST OF PLATES

INTRODUCTION

It is customary for authors to start a book, or end it, with a note in which they give credit to those who have helped in its production. In the case of this present book, which is one of illustrations in colour of a number of Venezuelan orchids, and some botanical notes concerning them, basic credit must go first to Nature who produced the flowers, and then to the makers of the faithful old Hasselblad camera with which the photographs were taken and also to those who invented colour photography and produced the Kodak and Agfa film with which the work was done.

For botanical knowledge we owe much to Dr. R.L. Dressler whose book "The Orchids", published by Harvard University Press, has been a valuable source of information, and above all to Dr. L.A. Garay, Curator of the Oakes Ames Orchid Herbarium of Harvard University, who has been our guide and mentor in orchid matters for the past 25 years. And finally, we owe this book to the skill and encouragement of Sr. Ernesto Armitano, the publisher, a perfectionist who finds it difficult to produce anything but the best quality in colour printing.

ORCHIDS

The Orchid Family, or *Orchidaceae* if you wish to be technical, is a member of a higher grouping called an Order, and is so specialized that some taxonomists place it in an Order called *Microspermae* on account of its very tiny seeds, where it has no companions other than two small and obscure families; yet other taxonomists place it even more exclusively in an Order known as *Orchidales* which contains just orchids and nothing else.

The Orchid Family competes with another great family, the *Compositae*, for the honour of Largest Family in the Vegetable Kingdom, containing from 15000 to 35000 species depending on who does the counting.

Few names of plant families have more *carisma* than that of the Orchids. Just the word *Orchid* seems to be enough to conjure up visions of expensive corsages of large and beautiful flowers, but for those who are lucky enough to live in Venezuela the word means much more than this, and there is nowadays a very extensive interest in *all* 'wild' orchids regardless of their size or where they come from. This is well demonstrated by a glance at the list of currently available books on orchids issued by an outfit in Michigan which specializes in this subject. This 'catalogue' lists no less than 174 items priced at not more than $50 each (including mail charges), resulting in an average of less than $25 per book - which in these days of inflation is by no means expensive.

With so much interest being shown by orchid *aficionados*, one could expect a fairly intimate knowledge of the Orchid Family to be quite common, yet even in Venezuela, where more than a thousand species grow wild in field or forest, there are many false characteristics attributed to orchids in the public mind. So it is perhaps wise to state right away that

1. Apart from the psychological effect on the lady who receives a corsage of orchids from her beau, the orchid has no proven aphrodisiac properties, and very limited medicinal value, if any.
2. They eat neither insect nor man.
3. They are not delicate: by comparison with most other plants they are tough.
4. They are not parasitic.
5. There is no truly black orchid.
6. No single wild orchid will make you a fortune, though a lucky hybrid bred in captivity might.

But besides questions of the 'How to grow' type, one not infrequently also hears questions as to just *what* is an orchid, or *where* do orchids grow and so on. To answer such queries in full botanical detail would take excessive space in this book, and would in any case be beyond our capacity. But within the limits of our non-technical background we can provide some answers to the main questions we are asked, using Venezuelan species as examples wherever possible.

The first such question is WHAT IS AN ORCHID? - in other words, what distinguishes an orchid from all other plants?

To this there is no simple, single reply, and it can only be answered by listing the many strange features that all together set the orchids apart. These features, very briefly and leaving out certain important but abstruse botanical details, are as follows:

A. The stamens and pistil which are separate sexual elements of flowering plants, are in the orchid fused (at least partly) into a single central element called the *Column*.

REPRESENTATIVE COLUMNS OF CERTAIN ORCHIDS

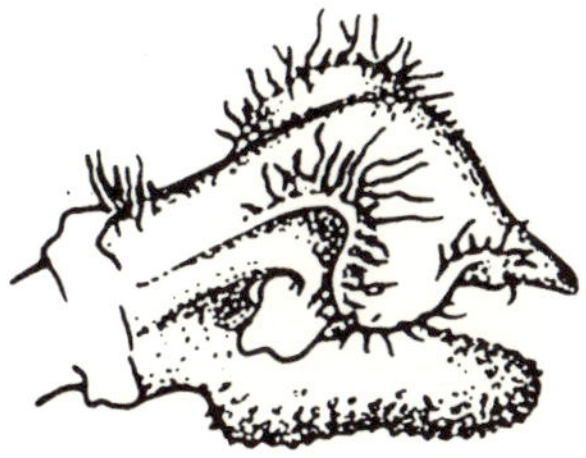

Phragmipedium lindleyanum

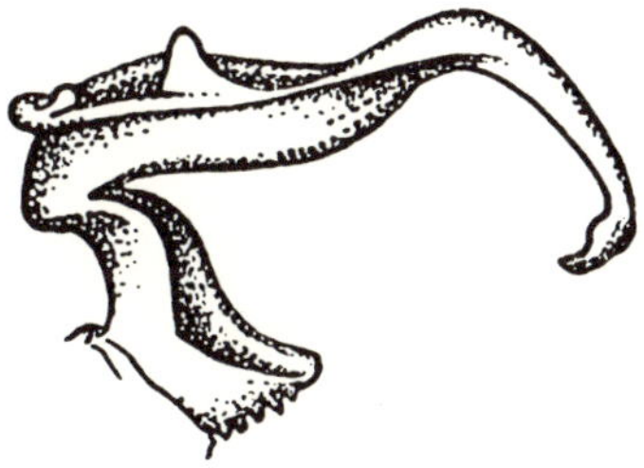

Ornithocephalus bicornis

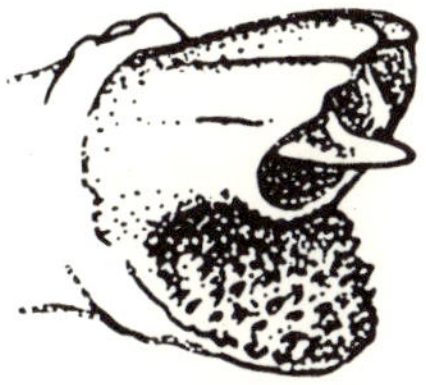

Pleurothallis linguifera

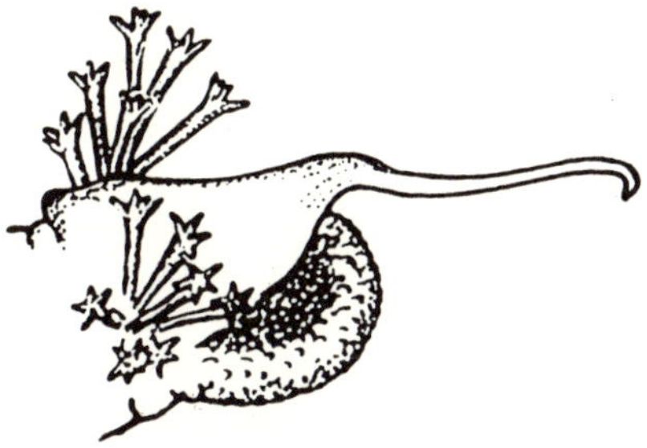

Stellilabium pogonostalix

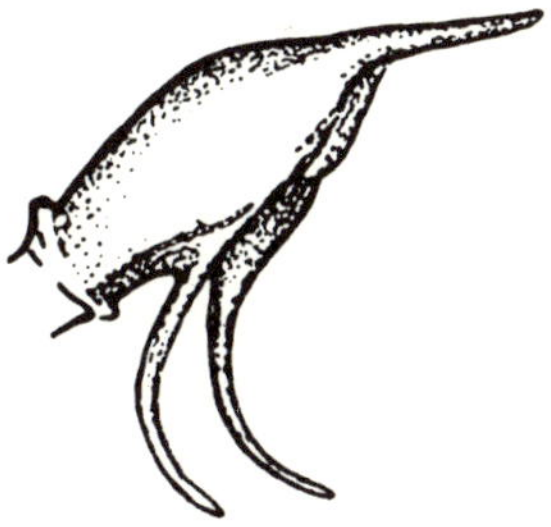

Catasetum sanguineum

Bulbophyllum meridense

Habenaria monorrhiza

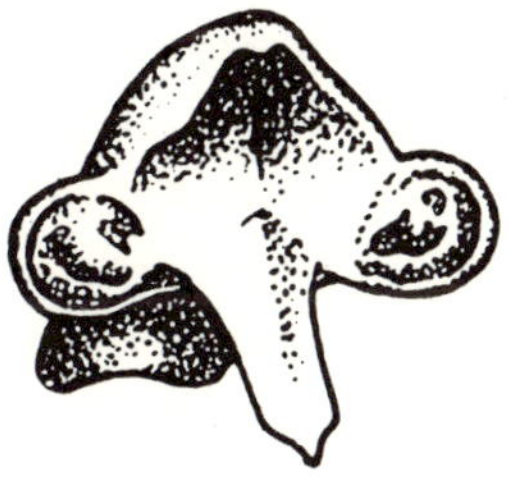

Stelis cucullata

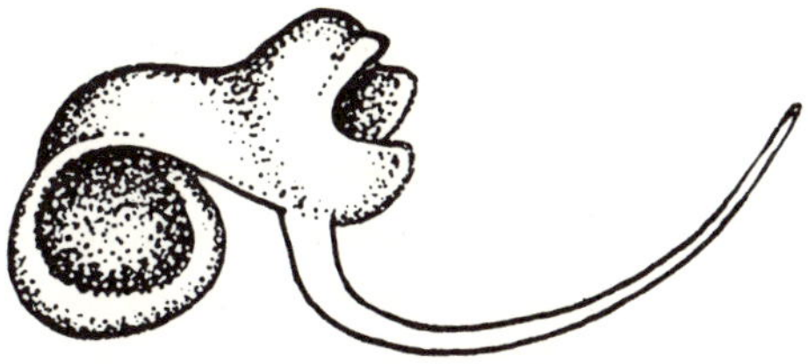

Dipteranthus planifolius

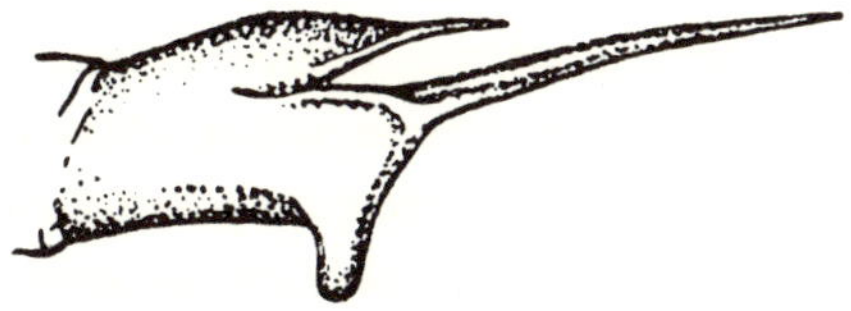

Spiranthes orchioides

Cryptarrhena lunata

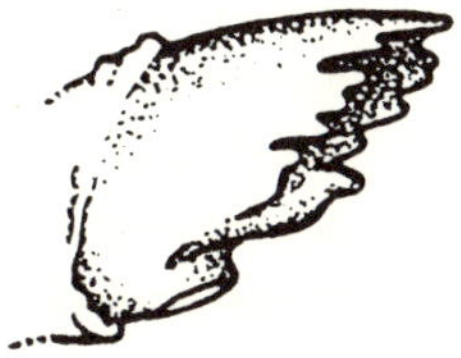

Kegeliella houtteana

While this element is sometimes 'columnar' in shape, it often looks most *un*like a column, as witness the columns of 12 Venezuelan orchids illustrated in the attached drawing:

Phragmipedium lindleyanum	*Ornithocephalus bicornis*	*Pleurothallis linguifera*
Stellilabium pogonostalix	*Catasetum sanguineum*	*Bulbophyllum meridense*
Habenaria monorrhiza	*Stelis cucullata*	*Dipteranthus planifolius*
Spiranthes orchioides	*Cryptarrhena lunata*	*Kegeliella houtteana*

B. The seeds are truly minute, and scientifically conducted experiments have shown that they can weigh as little as 3,000,000,000 to a kg. A check on the seed content in one capsule of the common Venezuelan orchid, *Cycnoches chlorochilon*, illustrated in this book, showed the figure of 3,770,000 seeds - and it is worth remembering that a single plant of this species can produce several such capsules each season. The seeds come fitted with an external coat of loose membranaceous material that greatly increases their buoyancy in the air and enables them to be dispersed by air movements over very long distances.

C. The flowers have an outer whorl of three sepals, often very colourful, but except in freak cases the inner whorl consists, not of the three petals that would be expected, but of two normal petals and a third one that has disguised itself by becoming a *labellum* (more commonly called a *lip*) which may be, and usually is, utterly different. The sepals may also appear in separated form or joined together in varying ways: when the two lateral sepals are so joined the united lamina is usually called a *synsepal*.

D. The flower in the bud starts in the 'primordial' attitude, with the lip at the top in a 'prone' position, but the pedicel twists during development so that the lip, when the flower opens, is 'supine' and is thus in the most favourable position, at the bottom of the flower, to expose its attraction not only to orchid lovers but more importantly to pollinating insects. This 180° twisting action is called *resupination*, but there are some orchids, such as certain species of the genus *Malaxis*, that are born with the lip in the natural primordial attitude but then have the quite extraordinary (and to us humans ridiculous) habit of twisting through the full 360° so that the lip once more comes back to the primordial prone position. At times one wonders whether to call the orchid a miracle of intelligence or stupidity!

E. What in nearly all other flowers would be loose pollen grains, in the orchid are stuck together in lumps of varied consistency known as *pollinia*. These pollinia are usually enclosed in a protective cap called an *anther* and connected by a *stipe* or *caudicle* to a sticky element known as a *viscid disc*, though often not at all disc-like in shape. The stipe is found in many shapes and sizes and is connected to the viscid disc by a membrane covering the side of the latter. Between the stigmatic surface and the pollinia there is generally a flap-like and most important element called the *rostellum*.

In conclusion, therefore, one can say that while other flowers may have one or more of the above features, only the orchid has them all.

But the orchid is a slippery opponent, difficult to pin to the mat, so it is worth mentioning that while, as proven by the column, the orchid is basically a bisexual flower, there are some genera, such as *Catasetum*, whose species sometimes produce unisexual flowers known as *pistillate* if they are female or *staminate* if they are male - sometimes one plant will have exclusively male or female flowers, but others may have a mixture of the two. Moreover, there are not only male and female forms but one can find, on a single raceme of a plant such as *Catasetum longifolium*, a whole gamut of intermediate forms starting with male and ending with female. One is tempted to ask, just how mixed up can an orchid get?

WHAT DOES AN ORCHID LOOK LIKE?

The Orchid Family is so large in number of species that it is not possible to say just what an *orchid flower* looks like. In size they can vary from that of the head of a very small dress-maker's pin in *Lepanthopsis steyermarkii* or *Platystele ornata* up to large *Cattleya* species with curiosities such as some *Phragmipedium* species which have slipper - or boat - shaped lips and extremely long, narrow, ribbon-like petals that dangle vertically down for lengths of up to 70 cm. In weirdness one can point to the miniature *Platystele johnstonii* with its absurdly long and slender apex on the dorsal sepal, or the flower of *Pleurothallis raymondii* which looks like a dangerous insect armed with petals ending in boxing gloves, or the extremely odd flowers of *Scaphosepalum verrucosum* that look like insects that have been taking LSD. Or the muscular *Porroglossum echidnum* that can trap an insect touching its lip, not to eat it but to make sure it fertilizes the plant; or the well-known *Oncidium papilio* which gets its name from its resemblance to a butterfly that has developed a third antenna; this flower is also an interesting 'freak' because while it seems to have three long antenna-like sepals and two normal petals, in fact two of the 'antennae' are petals and one is a sepal. As far as the complete inflorescence is concerned, one can quote *Dunstervillea mirabilis* at the 'short' end, with a peduncle and terminal raceme of 2 or 3 flowers, measuring 1 cm. overall, in contrast to the giant erect inflorescences of the terrestrial *Eulophia alta* standing as much as 3 m. tall, or the many epiphytic *Oncidiums* with long straggling inflorescences, such as *Oncidium volvox* which is at least 5m. long when straightened out.

But one could go on giving examples of the variability of flower and inflorescence almost indefinitely, so let us pass on to the form of the orchid plant as a whole.

Here again, the sheer variety of form prohibits any effort to explain what an orchid plant looks like. In size, a whole plant may vary from the above-mentioned *Dunstervillea mirabilis*, six of which could fit in a thimble, up to *Sobralia cattleya* with plants that may exceed 5m. in height. The size of a leaf varies from the 5mm. of *Chamelophyton kegelii* to the almost 1m.-long broad-bladed leaf of *Xylobium leontoglossum*; one can also mention the very extraordinary 'terete' leaf of *Scuticaria steelei* which is little thicker than a pencil but dangles from the branch of a tree for a length of 1.5m.

Some orchids grow in 'clumps', others have ever-lengthening rhizomes that clamber all over a tree. Some have no pseudobulbs, some have pseudobulbs almost too small to see, as in *Epidendrum manarae*, or cigar-shaped (fusiform) pseudobulbs, up to 50 cm. long, as in *Cyrtopodium glutiniferum*; while others, like an *Acineta* from the Andes that has not yet been clearly identified, have pseudobulbs the size of a large fist, 18 x 12 cm. If there is any clue to spotting an epiphytic orchid not yet in flower, it is by looking for its quite pale roots which spread over the trunk or branch of the tree and expose themselves to view often quite a distance from the orchid plant itself which may be well hidden.

This characteristic has led to orchids frequently being considered parasitic, as evidenced by their often being called *parásitos*, though they are not in any way parasitic as they take no nourishment from the tree but live on the chemical content of the dust and debris in the dirty rain water that washes down to them from higher up the tree.

MYCORRHIZA

A very intriguing question may have occurred to readers who have noted the remarks about 'millions of seeds' that can be easily and widely dispersed by air movements: this, combined with the Orchid Family's ability to find a home almost anywhere in the world, may make one wonder why the world has not already been smothered in orchids. The answer to this is that Nature, left to Itself, has its own ways of avoiding excesses of this sort, and in the case of the orchid it uses the very smallness of the seed itself to achieve this. A tiny seed, if it is to be airborne, cannot afford to travel encumbered with food supplies with which to start life elsewhere, and few would survive were it not for a complicated trick of Nature that helps it begin this new life with the aid of a normally deadly enemy - a fungus known as a *mycorrhiza*. When the seed lands where conditions are otherwise favourable for germination, the first stages of its precarious existence may start in the absence of this fungus, but most seedlings are unable to continue living without its help. Typically, the fungal hyphae enter the seed and penetrate the germinating embryo: in many cases the fungus kills the orchid seed, while in others the orchid kills the fungus, after which the orchid also dies for lack of food! But if the proper balance is established, the seedling is able to obtain enough food from the fungus to continue growing until it develops leaves, after which photosynthesis starts it on its own permanent way of life. It is thus clear that the orchid seed begins to grow in a very touch-and-go manner, with a high proportion failing to survive and thus the world is not swamped with orchids.

ORCHID TRICKS AND MECHANISMS

In common with many other plant families, the orchid relies on insects for fertilization, but differs in the firmness of its determination to achieve cross-fertilization and at all costs to avoid the damaging effects of self-fertilization of one flower by itself. The column, with its pollinia, and the lip, are the main elements brought into play by the orchid in the design of 'mechanisms' developed to effect these opposing ends; designs that are presumably the product of millions of years of trial-and-error. It is wrong to credit these plants with human intelligence and ingenuity, but this will not prevent us from writing as if they had such qualities when they created the 'tricks and mechanisms' described below, and which are by no means the only ones.

The first problem to be tackled by the orchid was probably the prevention of casual contact between the pollinia and the stigma. This it does by de-sexing one of the stigmas incorporated in the original column and causing the end to develop into a *rostellum*, to form a barrier between the pollinia and the active stigma. In an unhealthy plant or in some freak ones, this barrier may weaken and allow contact between pollinia and stigma, whereupon self-fertilization will take place, but the resulting 'incestuous' seedling is nearly always infertile or in some way defective.

The desired cross-fertilization takes place when the right kind of insect visits the plant and, attracted by its scent, colour, nectar or other lure, enters the flower, touches the viscid disc and thus goes off with the pollinia attached to it, in the process knocking off the anther 'cap' that previously protected the pollinia. The wrong kind of insect would be one to which the orchid denies entry. The means by which the orchid selects which insects are 'right' and which are 'wrong' are extremely ingenious and by themselves would need lengthy descriptions for which there is no space in this book. But some space must be devoted to very brief explanations of at least a few of the devices or 'tricks' invented by Nature to aid cross-fertilization while shunning self-fertilization like the plague.

Trick N° 1 This trick is so simple that it is surprising that the orchids ever bothered to invent other and far more complicated ones. In this trick the rostellum depresses itself when the pollinia are removed, and thus blocks access to the stigma for the same insect or others, until a prudent interval of time has passed, during which the insect with its load of pollinia will have gone elsewhere. When this period is over, the rostellum rises again, ready to welcome a visit from another insect loaded with pollinia from another flower, which will thus cause the required cross-fertilization.

Trick N° 2 This trick is used by the small Venezuelan orchid called *Porroglossum echidnum*, which has a sensitive lip with a muscular hinge at A connecting it to the column foot B. When an insect lands on the open lip and touches the sensitive spot C, the hinge contracts and the lip sweeps up with fair rapidity, to imprison the insect between lip and column in such a way that it can only escape by first passing the stigma D where it deposits its load of pollinia from an earlier flower, and then picks up a new load to take elsewhere. While so imprisoned it cannot back up to put the new load of pollinia it has just robbed back on the stigmas (which would thus cause self-fertilization) and by the time the lip opens up again the insect may be too exhausted to try revisiting the same flower.

Trick N° 3 This involves the stipe attached to the pollinia. When the insect enters the flower and contacts the viscid disc, it removes the pollinia with the anther cap, whereupon the stipe immediately curls up in such a way as to keep the anther cap in place on the pollinia and the insect thus flies off with all this impedimenta on board. If the insect then makes a quick return to the flower it will be unable to effect fertilization, but after the usual lapse of time the stipe will dry up, uncurl, and release the anther cap, exposing the pollinia then ready to fertilize another flower, the insect having forgotten its intention to make a second pass at the original plant.

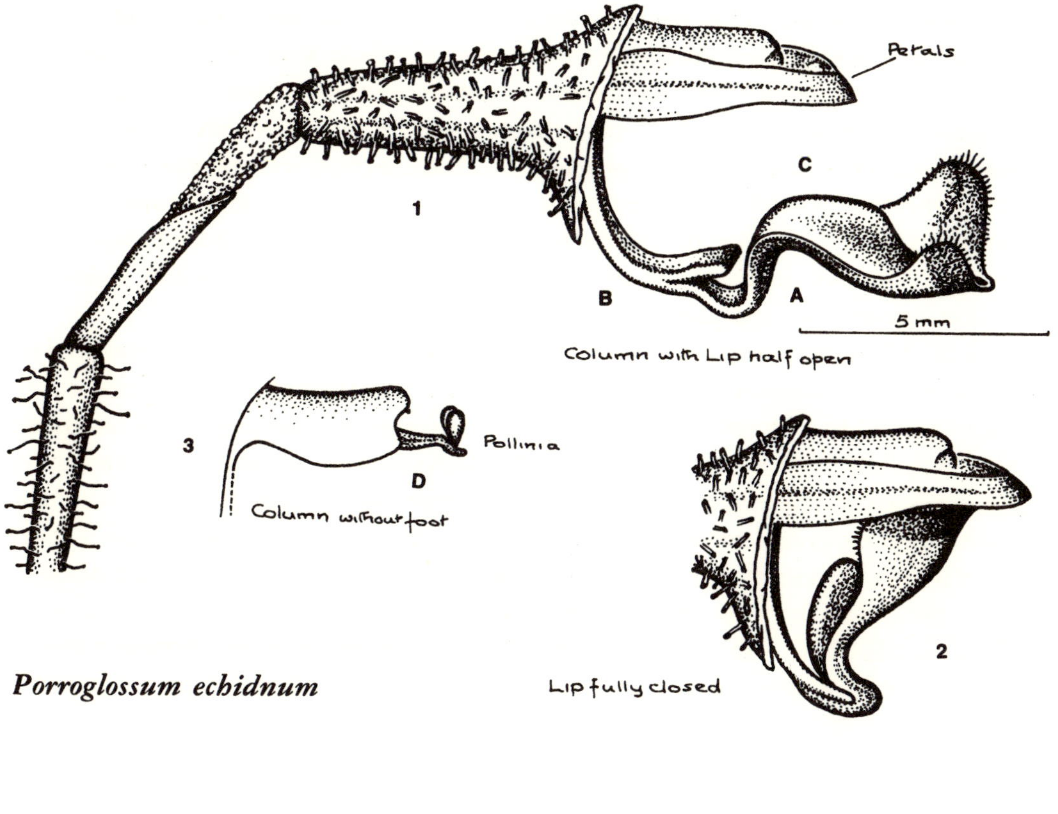

Porroglossum echidnum

Trick N° 4 This again involves the stipe. When the visiting insect contacts the viscid disc, it flies away with the pollinia in a position on its head or back in a poor attitude for contacting the stigma of the column from which it has just robbed the pollinia, so an immediate return to the same flower will fail to effect self-fertilization. But again, after a cetain time, the attitude of the stipe and pollinia will revert to the original position as the viscid disc dries and be ready to cross-fertilize a different plant.
Trick N° 5 This trick is based on a most complicated formation of the lip and column of the flower (in this case a *Coryanthes biflora*), which must have taken Nature untold ages to perfect. The lip of this, and some other *Coryanthes* species, is large and heavy, and hangs from the pedicel in the form of a sort of bucket (as shown in the attached figure), with the column hanging inside it. At the base (now the top) of the column there are two small glands, one on each side, that excrete a goodly amount of translucent liquid, drop by drop. The base of the bucket thus accumulates a noticeable quantity of this liquid, and the bees, attracted by some tasty material on the side near the rim, seem to get intoxicated with this and finally fall into the bucket. As they cannot climb the vertical walls they have to seek another way out if they are not to drown. This exit is in the front near the base, well hidden behind the apex of the column and so placed that in escaping the insect has to brush by the viscid disc and collect its load of pollinia, having first deposited on the stigma any pollinia it may already be carrying. It is most unlikely that the insect will try to repeat this process immediately and in any case scientific observations of this flower have shown

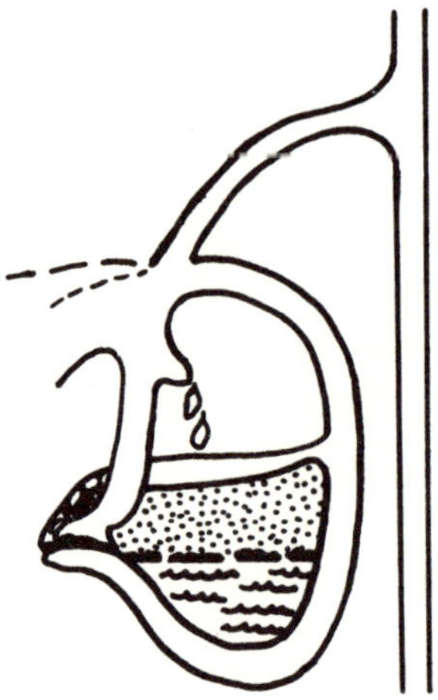

that sometimes there are so many bees assembled that there is a continual procession of them through the passage.

Trick Nº 6 For reasons known only to themselves, species of the genus *Catasetum* are not content with expecting the visiting insect to contact the viscid disc when it enters a plant, but take the extraordinary measure of shooting the pollinia and the viscid disc at the unsuspecting insect. In these orchids, the stipe is curled up round a convex rostellum in such a way that it forms a strong spring, held under considerable tension by a trigger formed by one or both of two long arms that extend down from the sides of the column. When an insect is large and strong enough to exert pressure on one of these arms the pollinia are catapulted with considerable force, enough to reach far out from the flower if they fail to make contact with the insect which has pulled the trigger. The real purpose of this dramatic mechanism is not clear unless it is to scare the insect away until it has recovered its wits again; since it is not likely to return, self-fertilization is prevented.

Trick Nº 7 We have left this trick for last as it is quite unbelievable how it ever came to be discovered by unthinking plants. At some very early stage of its development, a species of the orchid genus *Ophrys* found that a hairy burrowing insect called *Scolia ciliata*, which inhabited the same part of the Mediterranean, had the odd habit whereby the male insects came out of their burrows for a life in the open about a month before the female insects emerged for mating. The *Ophrys* orchid, by design or good management, not only happens to have a lip that looks like the insect but also opens its flowers just at the time when the lustful male insects are appearing, with the result that the male insect, deceived into thinking that it has found a wife, lands on the lip of the flower and does its best to mate. In the process the insect removes the orchid's pollinia and carries it to the next flower, thus fulfilling the orchid's wish for cross-fertilization. Self-fertilization is avoided because the insect, having (it thinks) dealt with one female obviously does not want to dally longer with her and so moves on to continue its conquests.

But incredible as this trick (known as *pseudo copulation*) may seem, what is even more incredible is that it has been discovered and used by other insects and other orchids in different parts of the world. ■

ACINETA ALTICOLA C. SCHWEINF.

Distribution: Venezuela, endemic. Can be found in the state of Bolívar, mainly altitudes from 1000 to over 2000m. on tepuis such as Auyantepui, Chimantatepui or Toronotepui.

As with all species in the genus *Acineta*, the inflorescence develops as a many-flowered pendent raceme

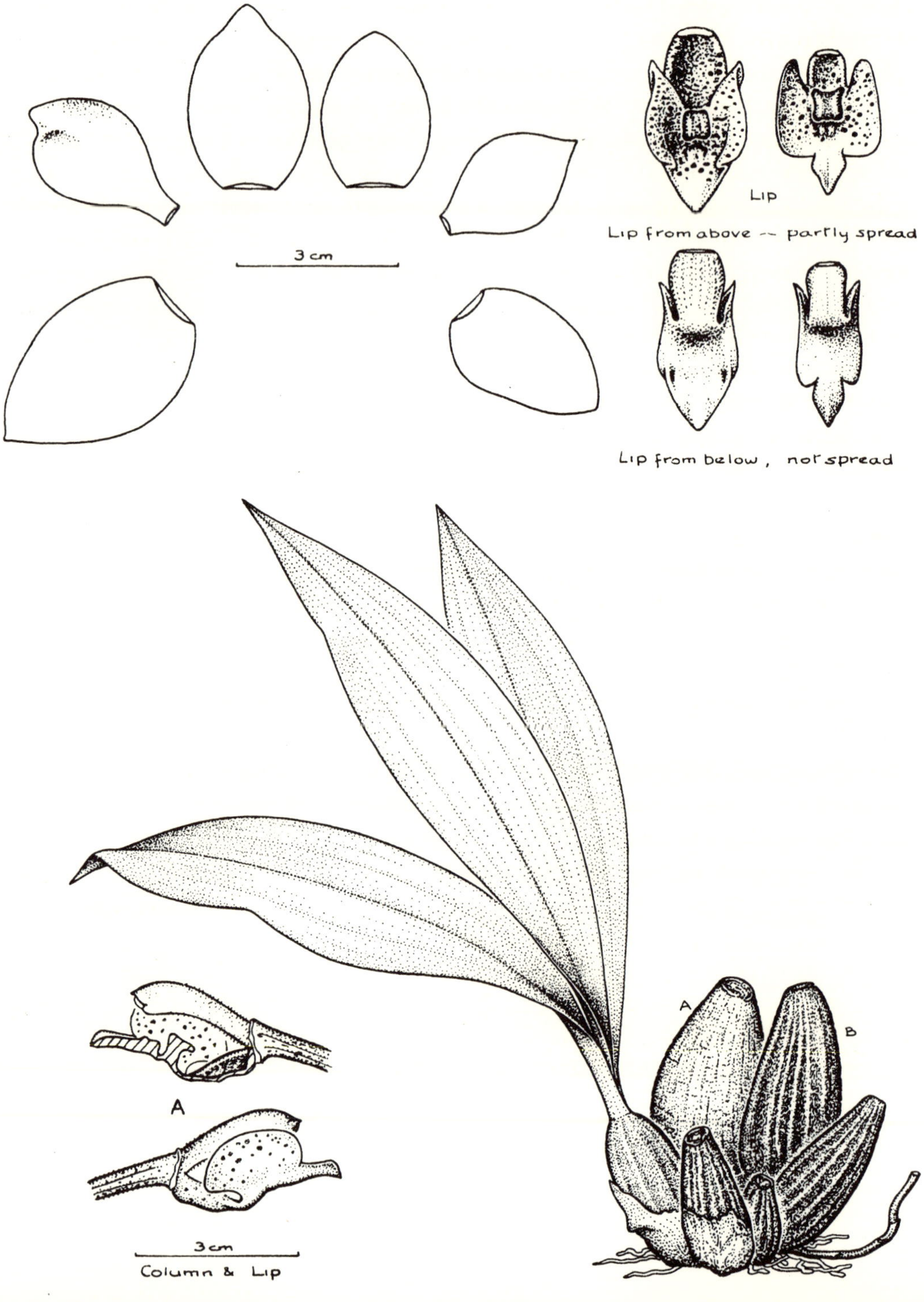

ACINETA ALTICOLA

ACINETA ERYTHROXANTHA

ACINETA ERYTHROXANTHA RCHB. F.

Synonym: *Acineta volteriana* Schltr.
Distribution: Venezuela and Colombia. In Venezuela it is known from near Sanare (Lara State) and from above Boconó (Trujillo State), where it grows epiphytically at about 1800m.

Lip sectioned

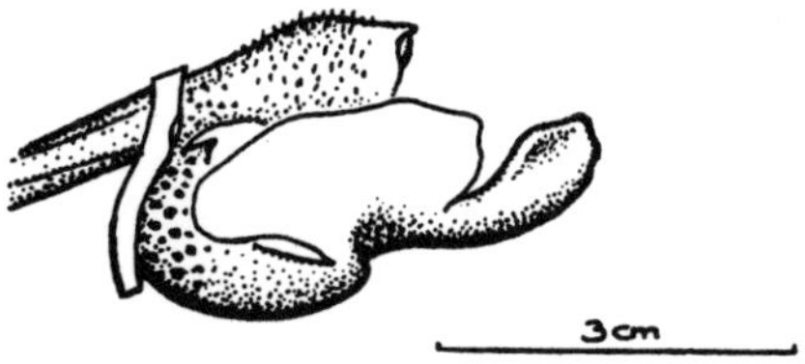

Column & Lip

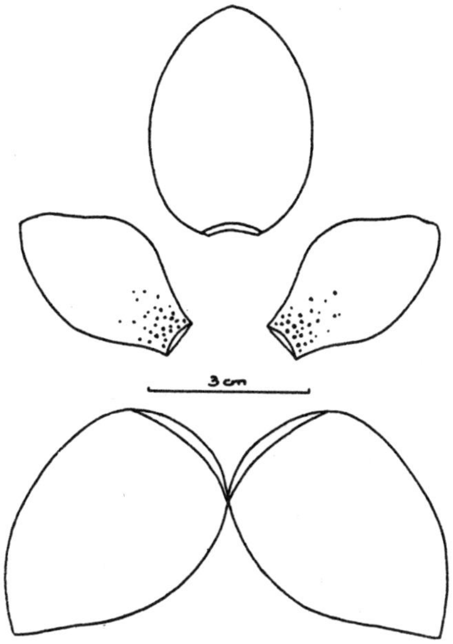

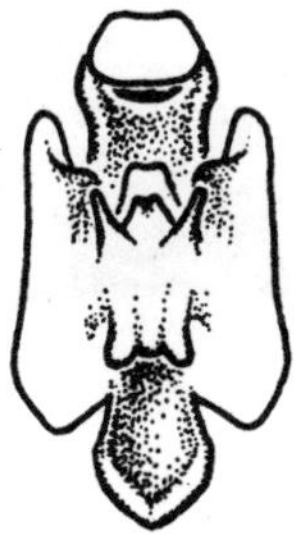

Lip
(Colouring not shown)

CATASETUM BARBATUM (LINDL.) LINDL.

Synonyms: This species is so variable in the form of its lip that nearly 20 names have been given to it, all of which are now listed as Synonyms of the name *C. barbatum*, at least until valid criteria have been found to separate them again. It was first named *Myanthus barbatus* by Lindley in 1835, but in 1844 he changed this to *Catasetum barbatum*.

Distribution: Venezuela, Guyana, Colombia, Ecuador, Peru and Brazil. In Venezuela it is widespread throughout the southern part of the country growing as an epiphyte but also in soil or on rock, at altitudes as low as almost sea level in the Federal Territory of Amazonas or Delta Amacuro or as high as 1000m. in the Gran Sabana of the state of Bolívar.

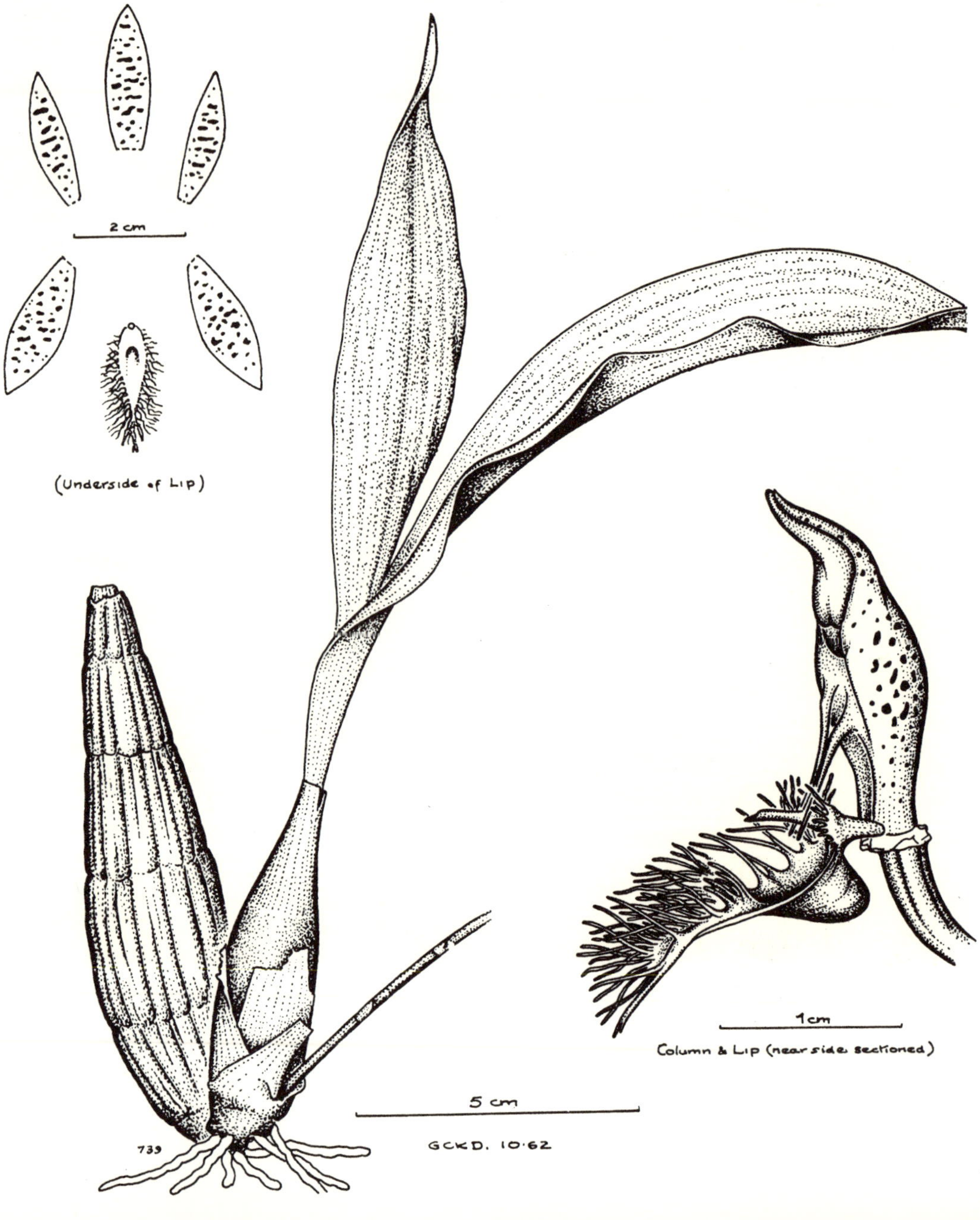

CATASETUM BARBATUM

ANGULOA CLOWESII

ANGULOA CLOWESII LINDL.

Synonym: *Anguloa clowesiana Williams*
Distribution: Venezuela and Colombia. In Venezuela grows epiphytically at altitudes around 1500 to 2000m. in the states of Mérida and Táchira.

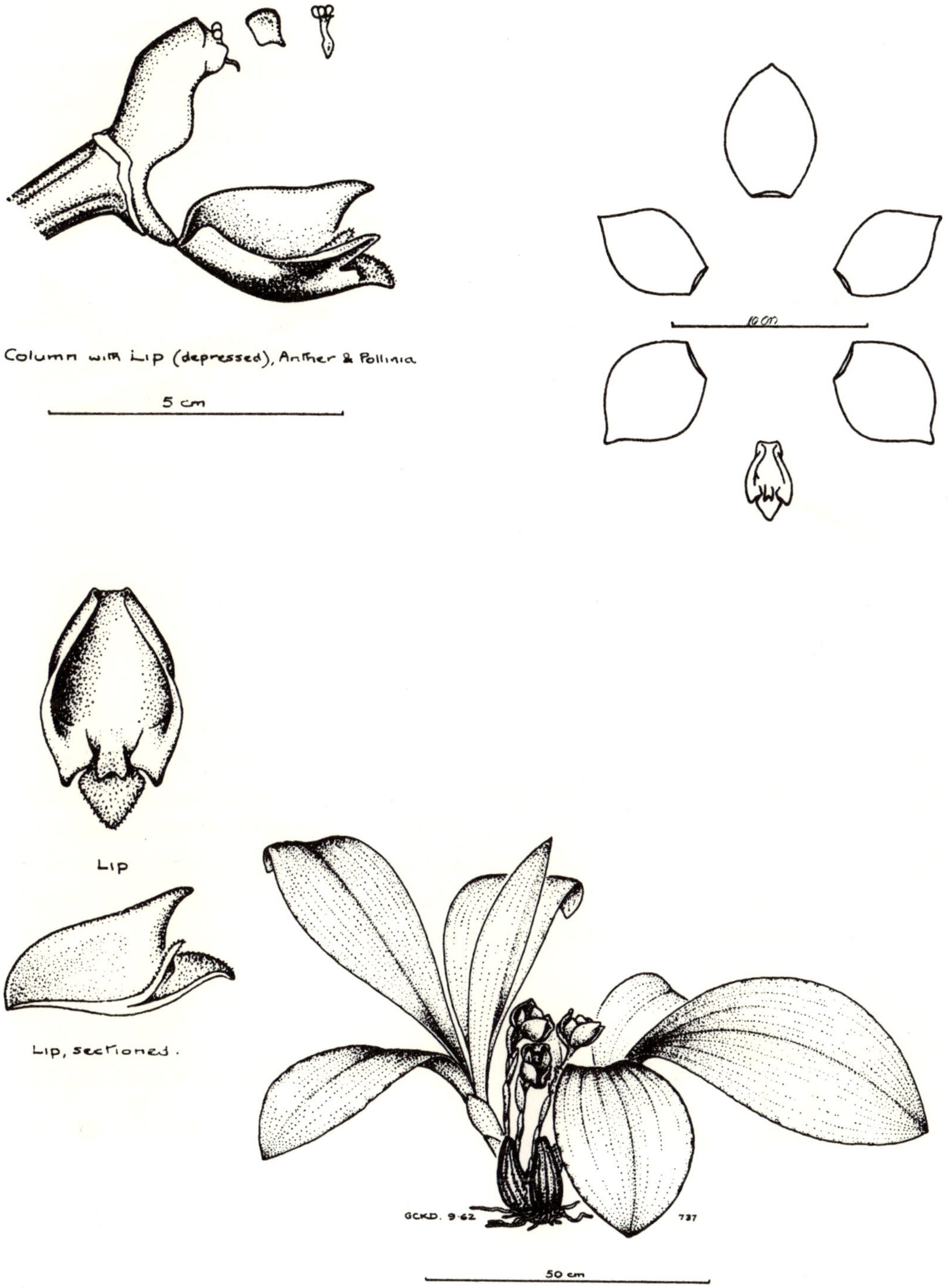

BRASSAVOLA NODOSA (L.) LINDL.

Synonyms: *Epidendrum nodosum* Linn.
Cymbidium nodosum Sw.
Bletia nodosa (L.) Rchb.f.
Bletia venosa (Lindl.) Rchb.f.
Brassavola grandiflora Lindl.
Brassavola venosa Lindl.

Distribution: Venezuela, the West Indies, Mexico to Panama and Venezuela to Peru. In Venezuela it is to be found on small trees, or even on cactus plants, from sea level to 1800m., in the states of Aragua, Carabobo, Falcón, Lara, Mérida and Zulia. It has subterete and fleshy leaves which are well suited to life in dry habitats.

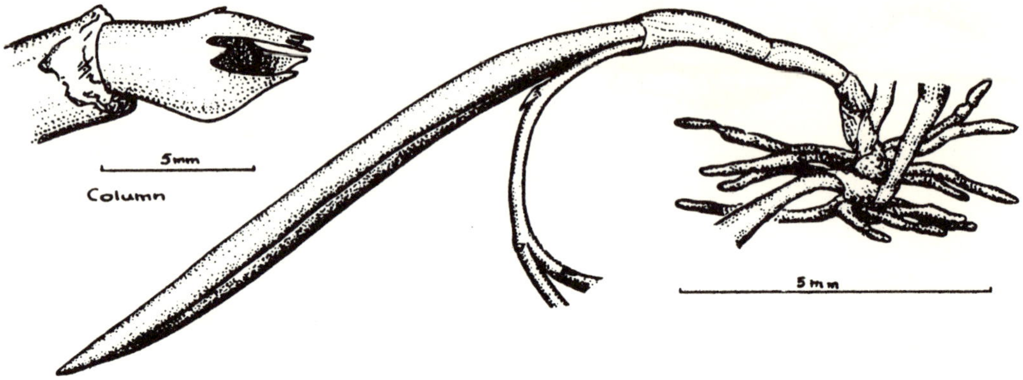

BRASSIA ANGUSTA LINDL.

Synonyms: *Brassia lawrenceana* var. *angusta* (Lindl.) Lindl.
Oncidium lawrenceanum var. *angustum* (Lindl.) Rchb.f.

Distribution: Venezuela, Guyana, Brazil. In Venezuela it occurs as an epiphyte in the forests of Bolívar state and the Federal Territory of Amazonas, from 140m. altitude on the Siapa River to 1200m. on the Parima Range bordering on Brazil. The spidery flowers, each almost 10cm. in size and growing in racemes of up to 30 flowers make an attractive display, but as the raceme is usually horizontal in attitude it does not make a good showing if surrounded by other plants.

BRASSAVOLA NODOSA

BRASSIA ANGUSTA

CATTLEYA MAXIMA

CATTLEYA PERCIVALIANA

CATTLEYA MAXIMA LINDL.

Synonym: *Epidendrum maximum. Rchb.f.*
Distribution: Venezuela, Ecuador, Peru. In Venezuela we have seen specimens of this species reported by reputable persons as having been found in the state of Bolívar near Kavanayén and near the border with Guyana on the Island of Anacoco.

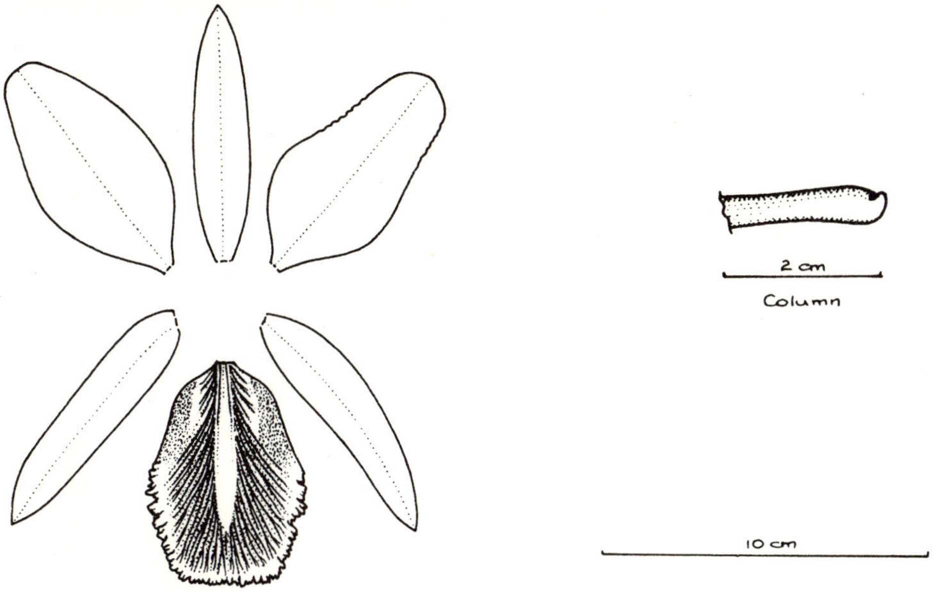

CATTLEYA PERCIVALIANA O'BRIEN

Synonyms: *Cattleya labiata var. percivaliana Rchb.f.*
Distribution: Endemic to Venezuela, where it grows in the Andean part of the country at some 1300 to 1700m. in the states of Mérida and Trujillo.
This is a very popular orchid with connoisseurs and has a number of color varieties, ranging from rose-purple throughout to all white or to semi-alba forms that have white sepals and petals and colored lips as illustrated here. The perfume of this species is unpleasant to some people and is very distinctive.

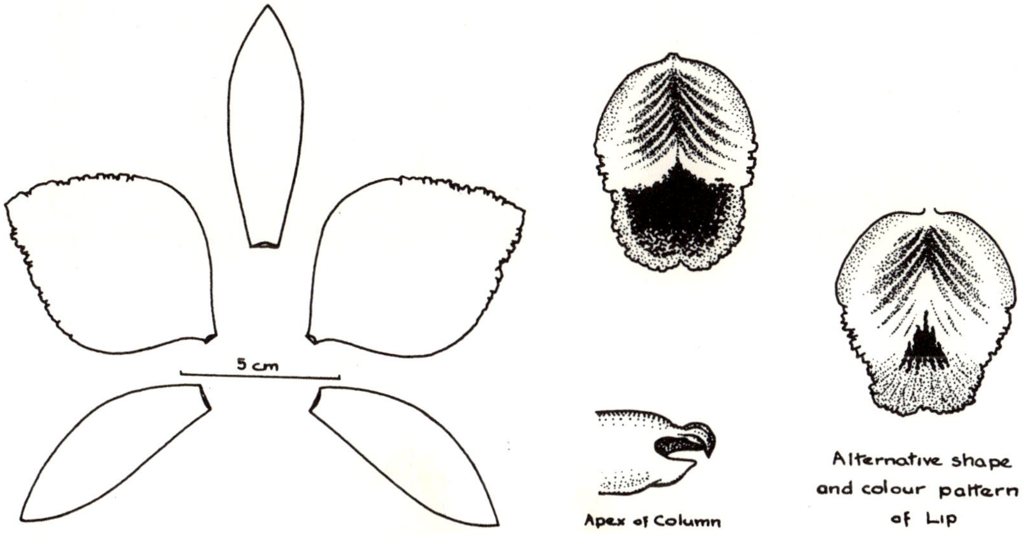

BIFRENARIA MAGUIREI C. SCHWEINF.

Distribution: Venezuela only, where it grows at altitudes of 1300-1500m. in the soil of the highlands of the Federal Territory of Amazonas such as Cerro Autana, C. Sipapo, C. Yapacana, C. Duida and the Aracamuni section of C. Avispa.
This is a most attractive orchid with erect stems to a metre tall, bearing 2 to 3 flowers 10cm. in size, and so far has been protected from over-collecting by the remoteness and inaccessibility of its habitat.

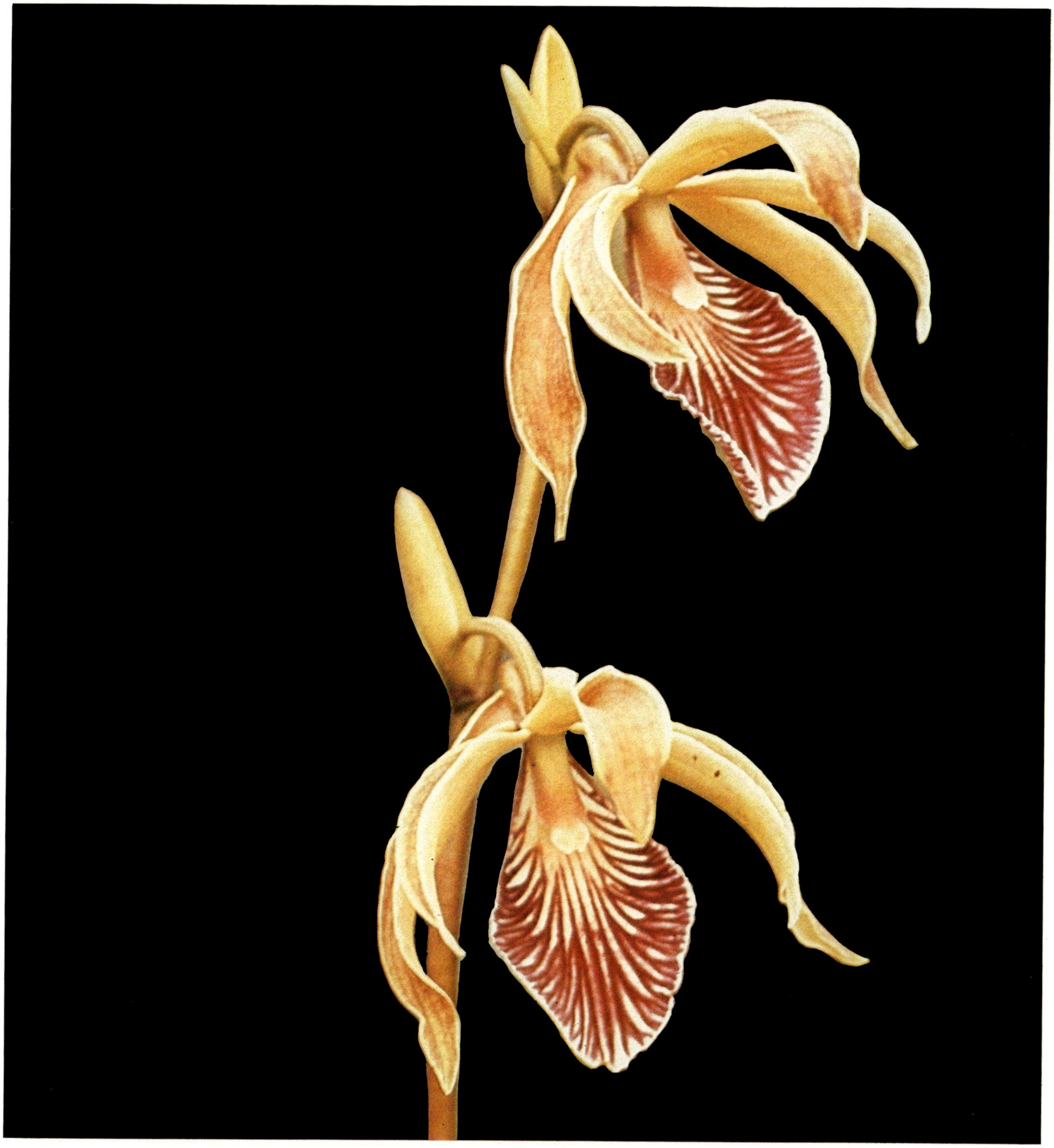

BIFRENARIA MAGUIREI

ANGULOA RUCKERI

ANGULOA RUCKERI LINDL.

Distribution: Venezuela, Colombia, Ecuador. In Venezuela it occurs as an epiphyte at altitudes from 900 to 1700m. in the states of Monagas, Mérida and Táchira. This robust species has rather dull flowers with an external coloration of green grading to an orange-brown, but the peasants near the well-known Guácharo caves in Monagas, near where this orchid can be found, have discovered the art of opening up the flower, without breaking off the rigid sepals and petals, to reveal their bright red colour, as shown here, and this greatly boosts their roadside sales!

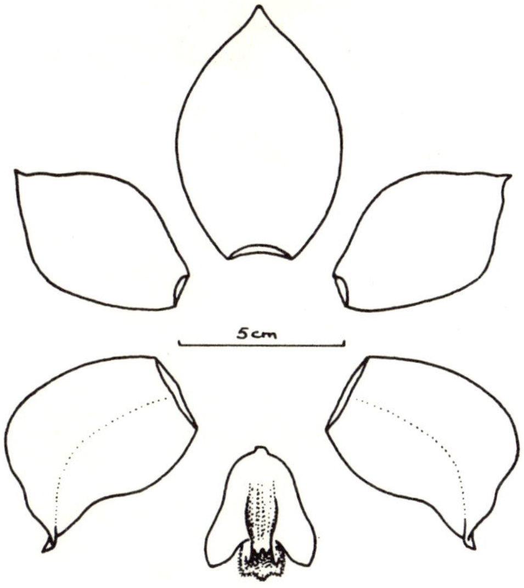

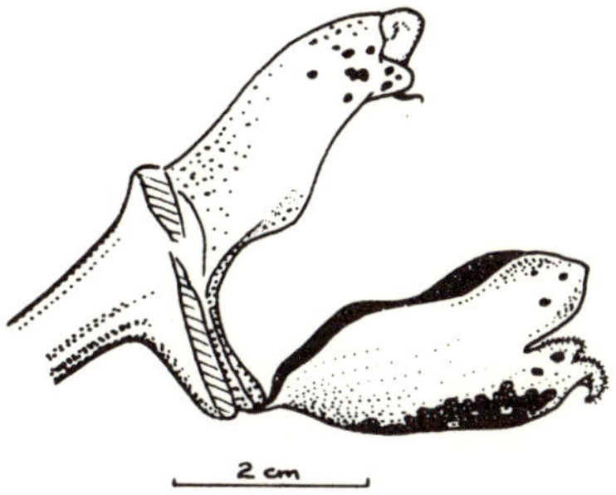

Column & Lip

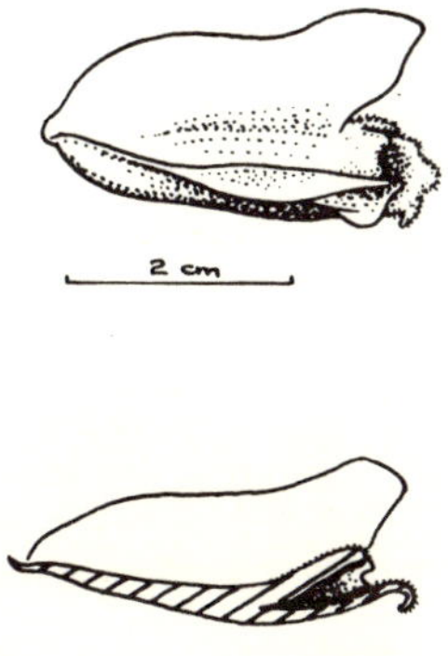

Lip, from above & sectioned

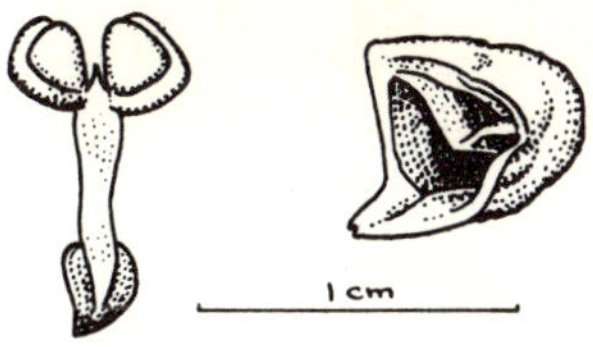

Pollinia Anther

BRACHIONIDIUM BREVICAUDATUM ROLFE

Distribution: Venezuela and Guyana. In Venezuela it is to be found in the Gran Sabana area of the state of Bolívar, at altitudes from 1200 to 2000m. where it grows epiphytically and also in the soil or at times on almost bare rock. It was first collected by McConnell and Quelch on Roraima in 1898 and published by Rolfe in 1901. For those who admire miniature orchids this is a rather rare and attractive species. The plant grows to about 10cm. in height and bears a relatively large flower.

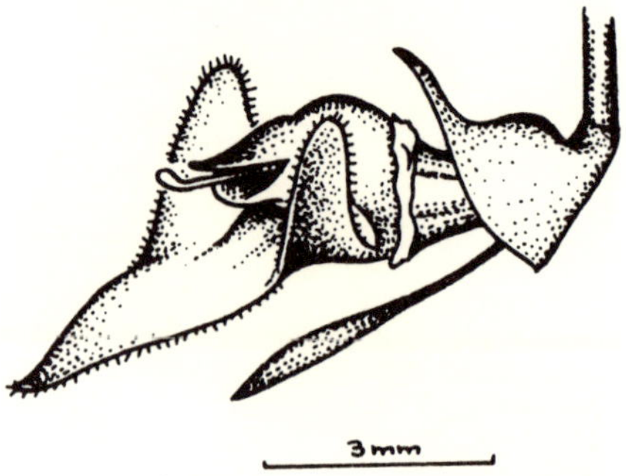

Column, Lip & Floral Bract (Inverted)

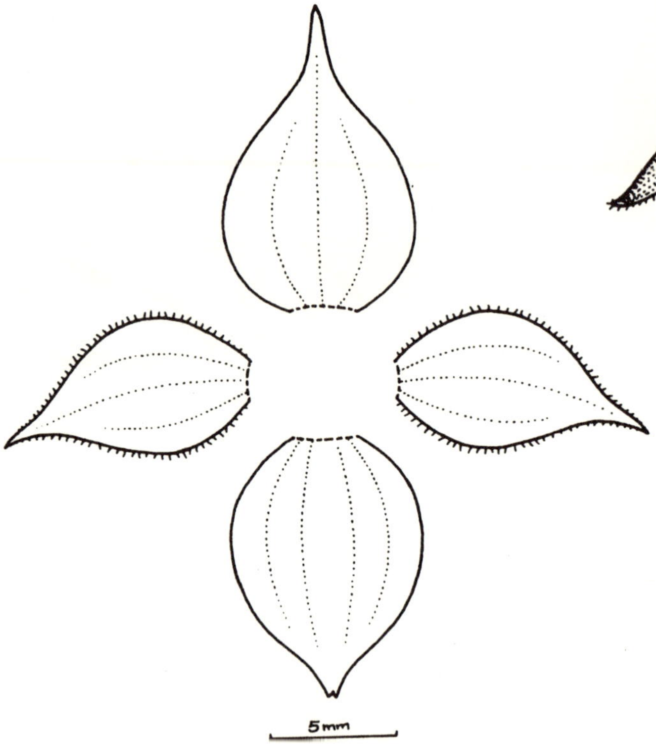

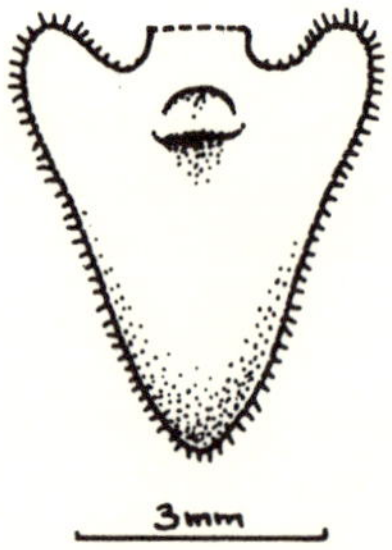

Lip

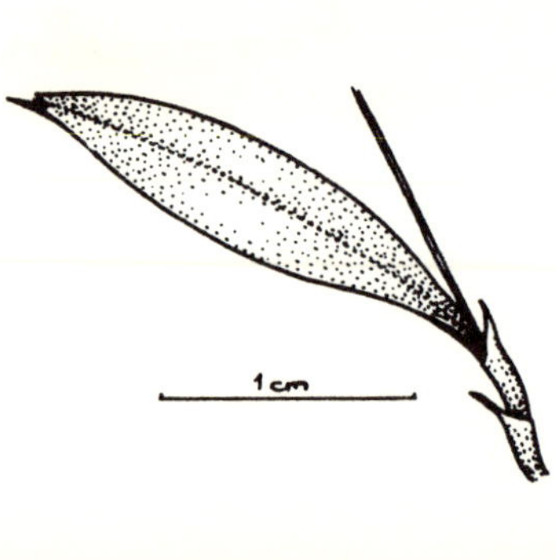

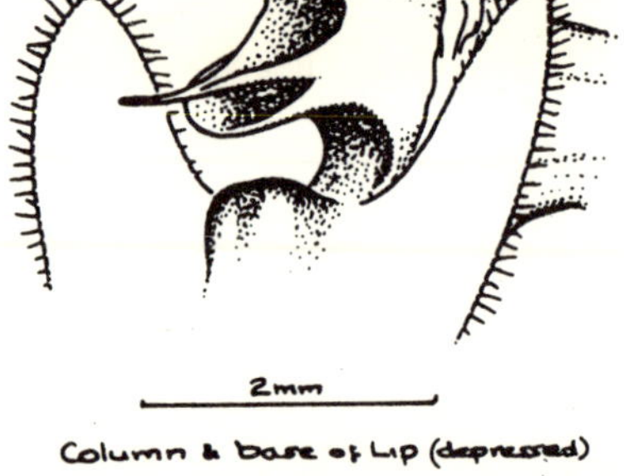

Column & base of Lip (depressed)

BRACHIONIDIUM BREVICAUDATUM

CHONDRORHYNCHA FLAVEOLA

CHONDRORHYNCHA FLAVEOLA (LINDEN & RCHB. F.) GARAY

Synonyms: *Zygopetalum flaveolum* Linden & Rchb.f.
Stenia fimbriata Linden & Rchb.f.
Kefersteinia flaveola (Linden & Rchb.f.) Schltr.

Distribution: Venezuela, Colombia, Ecuador and Peru. In Venezuela it grows epiphytically at altitudes from 1200 to 2200m. in the state of Lara and in the Perijá Range on the borders of Colombia in the state of Zulia.

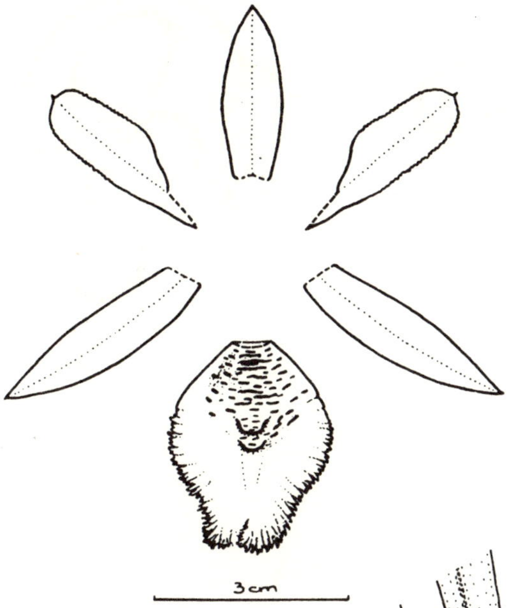

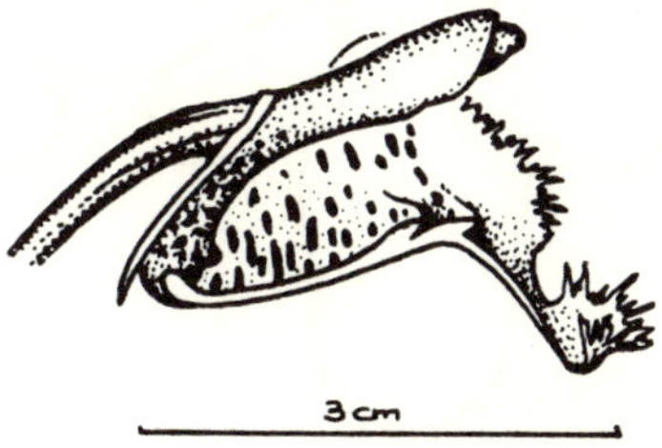

Column & Lip (sectioned)

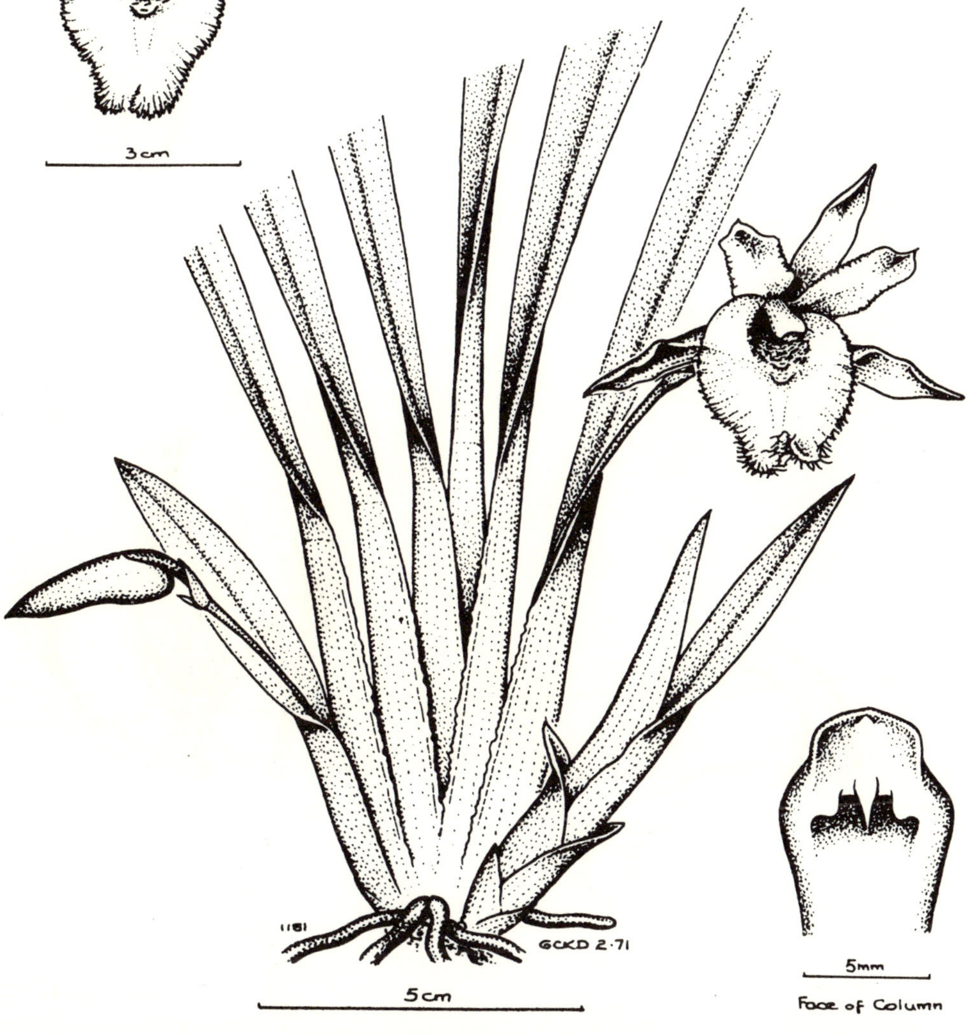

Face of Column

CHRYSOCYCNIS SCHLIMII LINDEN & RCHB.F.

Synonym: *Chrysocycnis trigonidii* Linden & Rchb.f.
Distribution: Venezuela and Colombia. In Venezuela this rather rare species has been found at about 2000m. in the western Andes. The plant starts with its initial roots in the soil, but the slender rhizome is scandent, with small pseudobulbs and roots occurring at intervals of some 20cm.

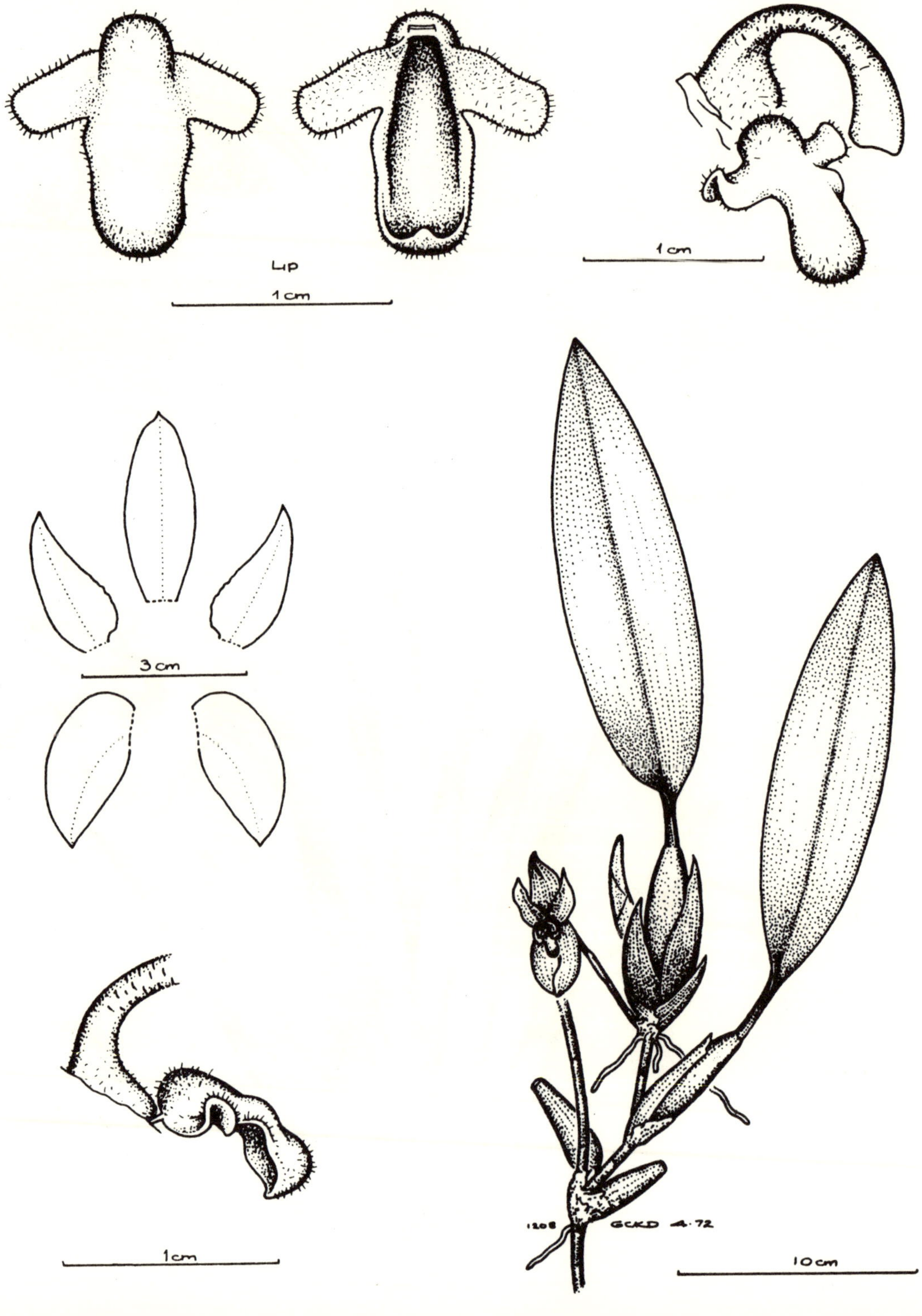

CHRYSOCYCNIS SCHLIMII

CATASETUM LONGIFOLIUM

CATASETUM LONGIFOLIUM LINDL.

Synonyms: *Monacanthus longifolia* (Lindl.) Hook.

Distribution: Venezuela, Guyana, Surinam. In Venezuela this species is limited to a small part of the Gran Sabana (Bolívar State) at an altitude of about 700m., not far from Sta Elena de Uairén, where it grows close under the crown of leaves at the top of *Mauritia flexuosa* (moriche) palms. The plant has the extraordinary habit of growing in an inverted fashion, with the long narrow leaves and the inflorescences hanging vertically down. This species is a fine example of the *dimorphism* to be found in a number of *catasetums*, the plants bearing flowers that may vary from pure male to pure female in form, and at times with a complete series of intermediate "hermaphroditic" forms developing on a single raceme.

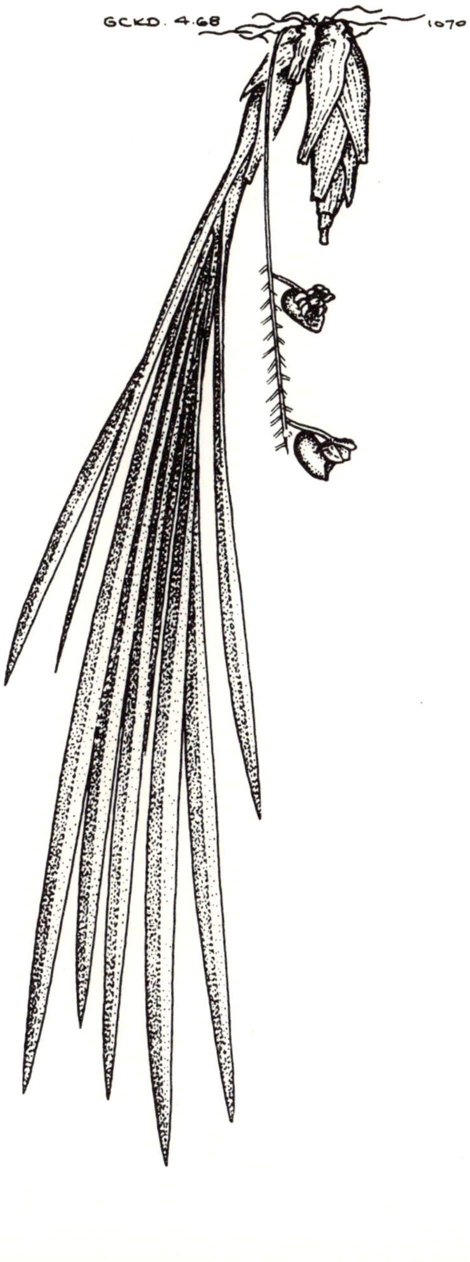

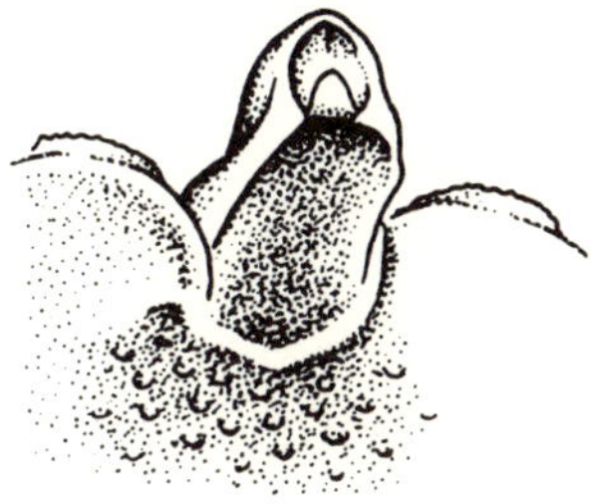

Face of Female Column

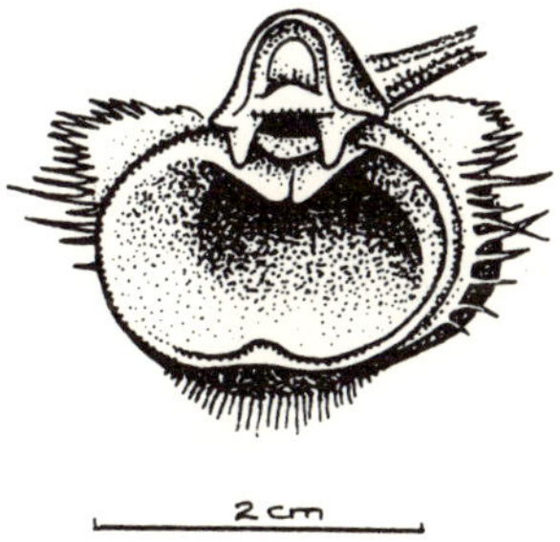

Male Flower

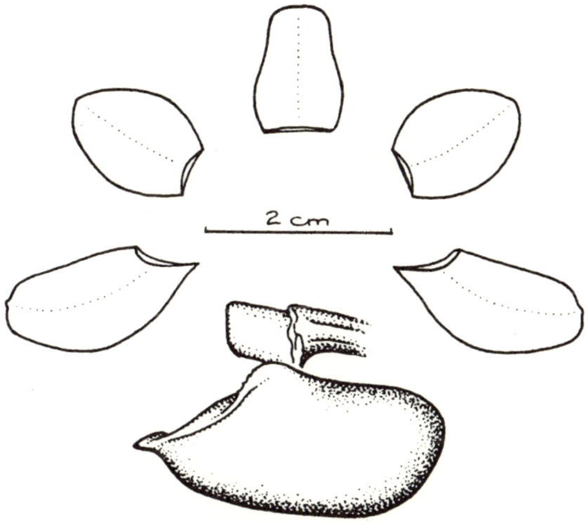

CATASETUM MACROCARPUM L.C. RICH EX KUNTH.

Synonyms: *Catasetum claveringhii* Lindl.
Catasetum tridentatum Hook.
Catasetum floribundum Hook.
Catasetum tricolor Hort. ex Planch.
Catasetum linguiferum Schltr.

Distribution: Venezuela, Trinidad, Guyana, Surinam and Brazil. In Venezuela it is a common epiphyte at altitudes from 500 to 1300m. in the region of Caracas and the Coastal Range, and also occurs in the state of Zulia.

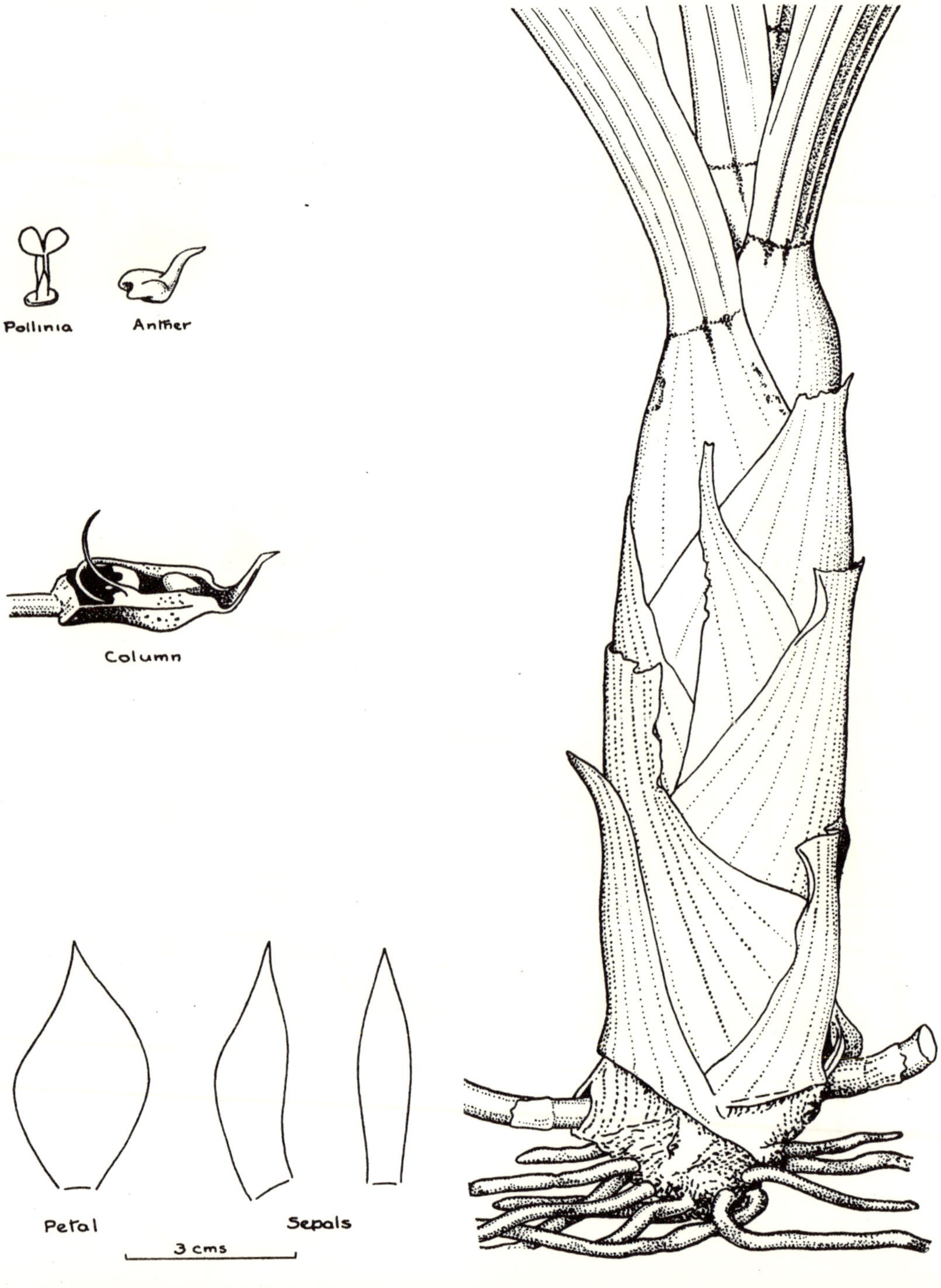

CATASETUM MACROCARPUM

CATASETUM PILEATUM

CATASETUM PILEATUM RCHB. F.

Synonym: *Catasetum bungerothii* N.E.Br.
Distribution: Venezuela, Trinidad and the Amazon region of Brazil, Colombia and Ecuador. In Venezuela it is found in the region of the Upper Orinoco in the Federal Territory of Amazonas, mostly at elevations not over 200m.; and it also grows in the state of Apure.
The normal color for this species is cream or light green with a lustre like mother of pearl which has given this orchid the local name of *Flor de Nacar*, under which name it was once selected as the the National Flower. There are many color forms of this species including one very fine one with a lip that is almost entirely red. A well developed pendent raceme can bear as many as twelve large flowers.

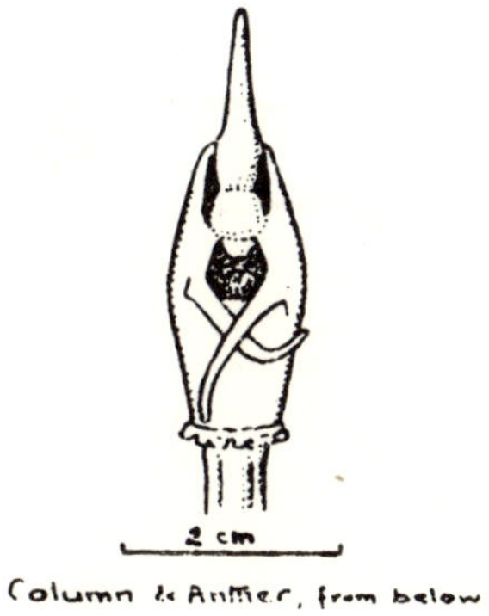

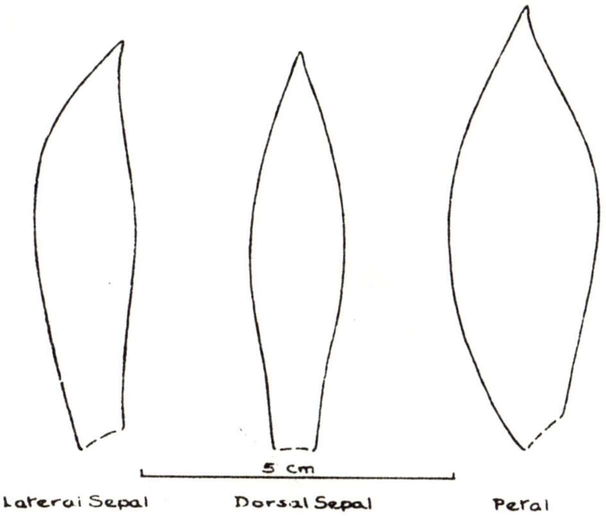

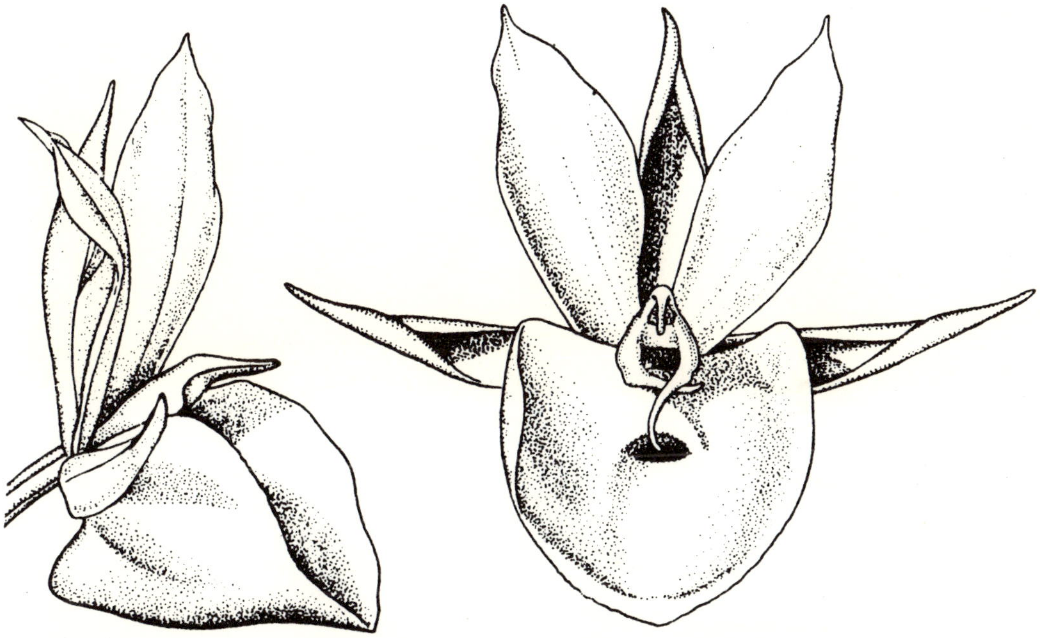

EPIDENDRUM FRAGRANS SW.

Synonyms: *Epidendrum cordatum* Vell.
Epidendrum aemulum Lindl.
Epidendrum ionoleucum Hoffmgg. ex. Rchb.f.
Epidendrum lambda Linden & Rchb.f.
plus 9 named varieties of E. *fragrans* itself.

Distribution: Widespread as an epiphyte throughout the American tropics and subtropics. In Venezuela it is also widespread from 50 to 1500m., and on rare occasions to over 2000m.

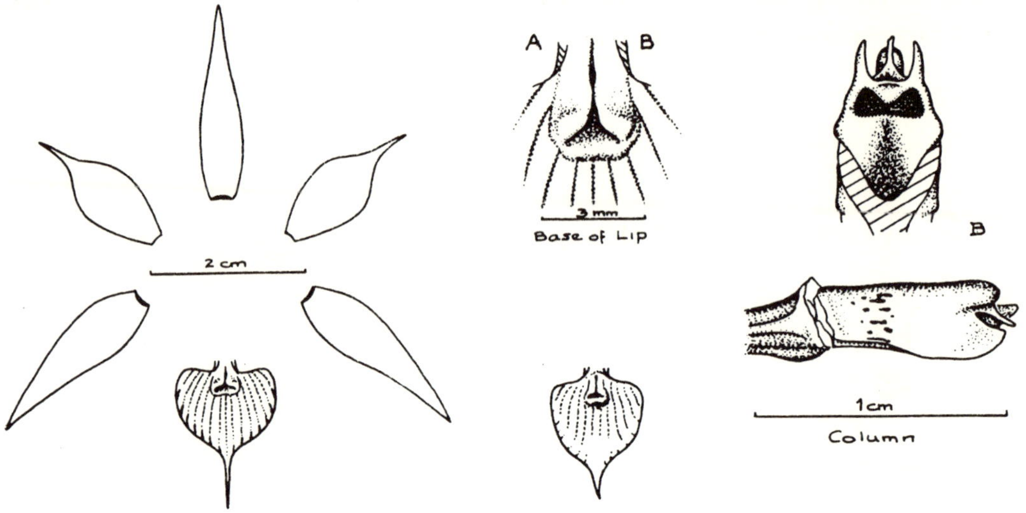

EPIDENDRUM GARCIANUM GARAY & DUNSTERV.

Distribution: Endemic to Venezuela where it was found in 1959 by the Caracas orchidist Carlos García E., growing as an epiphyte in the state of Falcón from 1200 to 1400m. It has subsequently also been found growing on rocks and low shrubs on Cerro Sta. Ana in the peninsula of Paraguaná in Falcón.

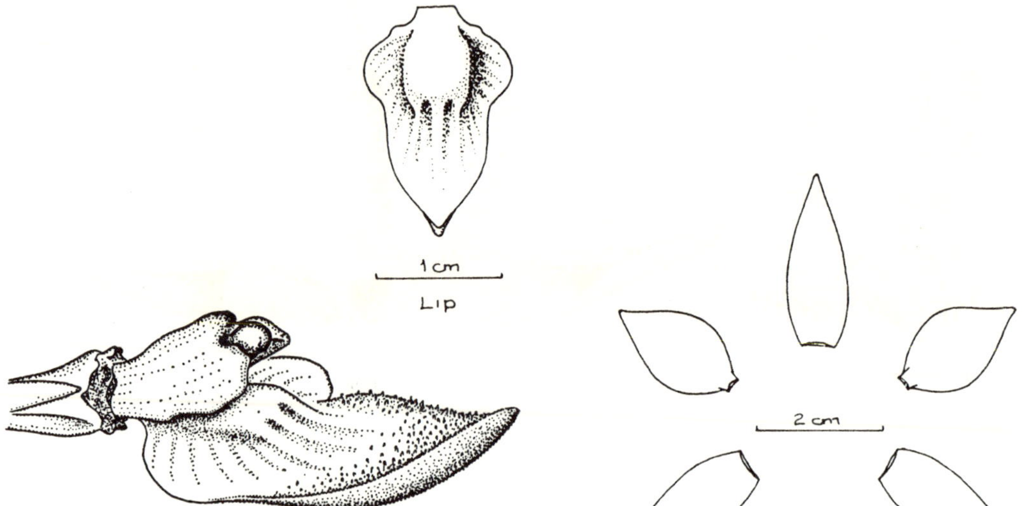

EPIDENDRUM FRAGRANS

EPIDENDRUM GARCIANUM

MENDONCELLA JORISIANA

MAXILLARIA NIGRESCENS

MENDONCELLA JORISIANA (ROLFE) HAWKES

Synonyms: *Zygopetalum jorisianum* Rolfe
Galeottia jorisiana (Rolfe) Schltr.

Distribution: Venezuela and Brazil. In Venezuela it grows as an epiphyte at altitudes from 1500m., and in the Federal Territory of Amazonas at 1800m. on Cerro de la Neblina.

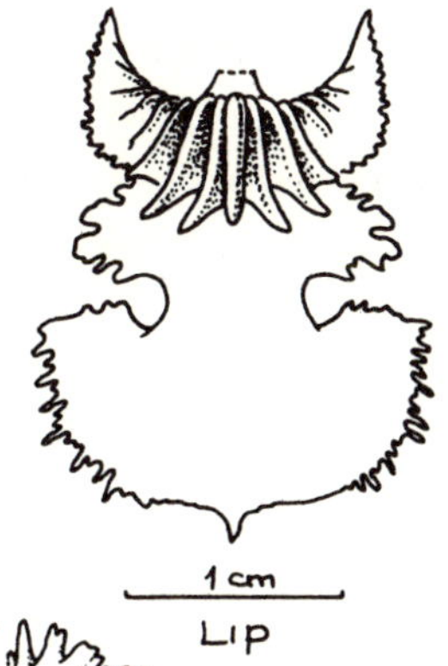

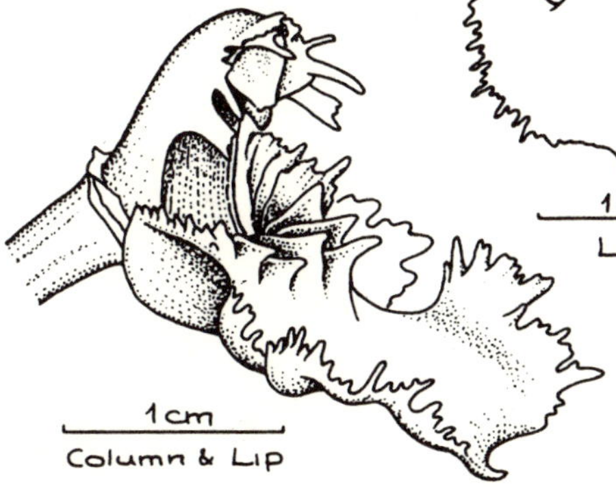

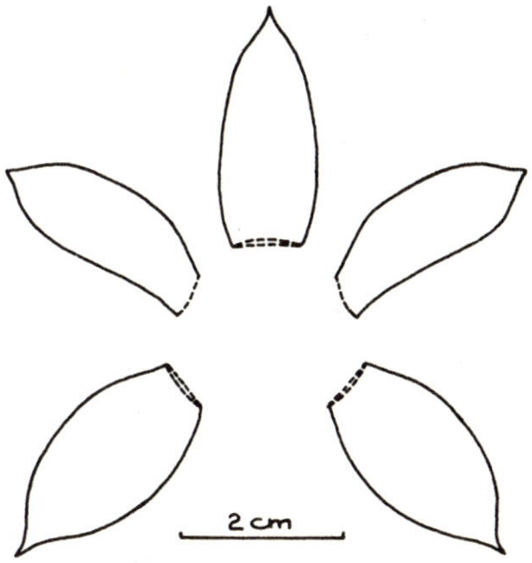

MAXILLARIA NIGRESCENS LINDL.

Synonym: *Maxillaria rubrofusca* Kl.

Distribution: Venezuela and Colombia. In Venezuela it can be found growing epiphytically at elevations from 1500 to 2600m. in the states of Trujillo, Mérida and Táchira. The natural attitude of its red sepals and petals gives the flowers the appearance of devils leaping for their prey!

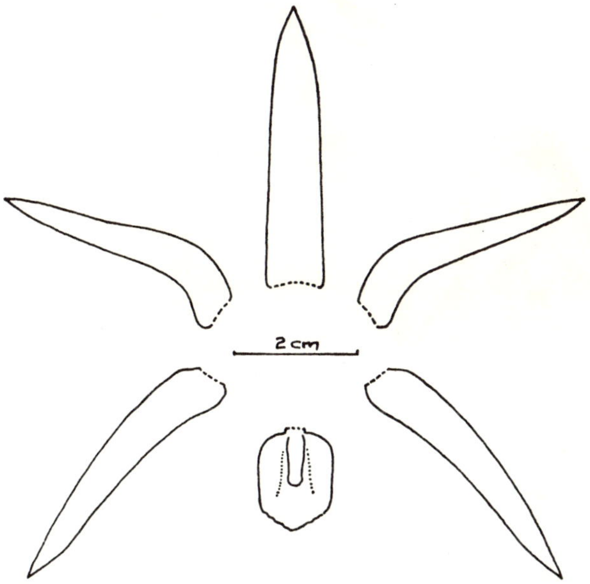

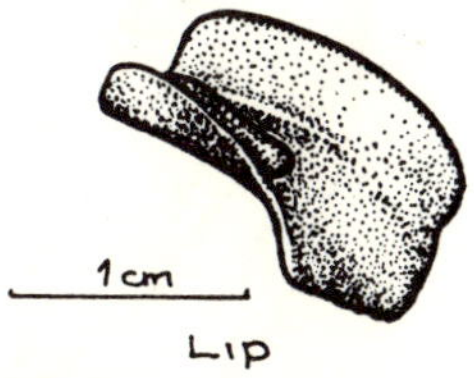

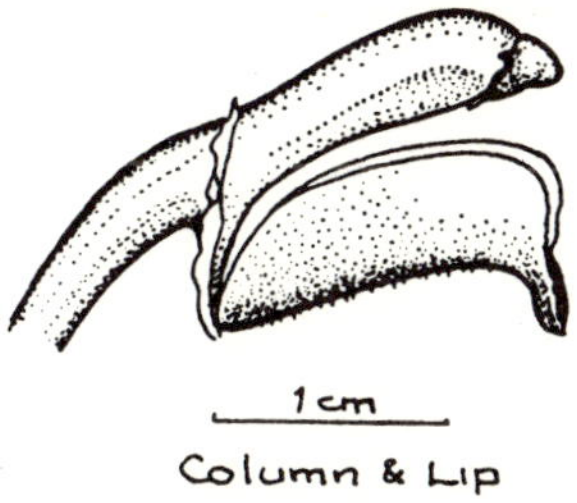

CATASETUM SPLENDENS COGN.

Synonyms: *Catasetum revolutum* Cogn.
Catasetum lindenii Cogn.
Catasetum lucianii Cogn.
Catasetum semiroseum Beck
Catasetum imperiale Lindl. & Cogn.
Catasetum cabrutae Schnee

Distribution: Venezuela and Brazil. In Venezuela it can be found growing epiphytically at low altitudes in the state of Apure but is principally known from the Upper Orinoco region of the Federal Territory of Amazonas. The inflorescence is arching, and the terminal raceme of twelve or more large flowers makes it a most impressive orchid.

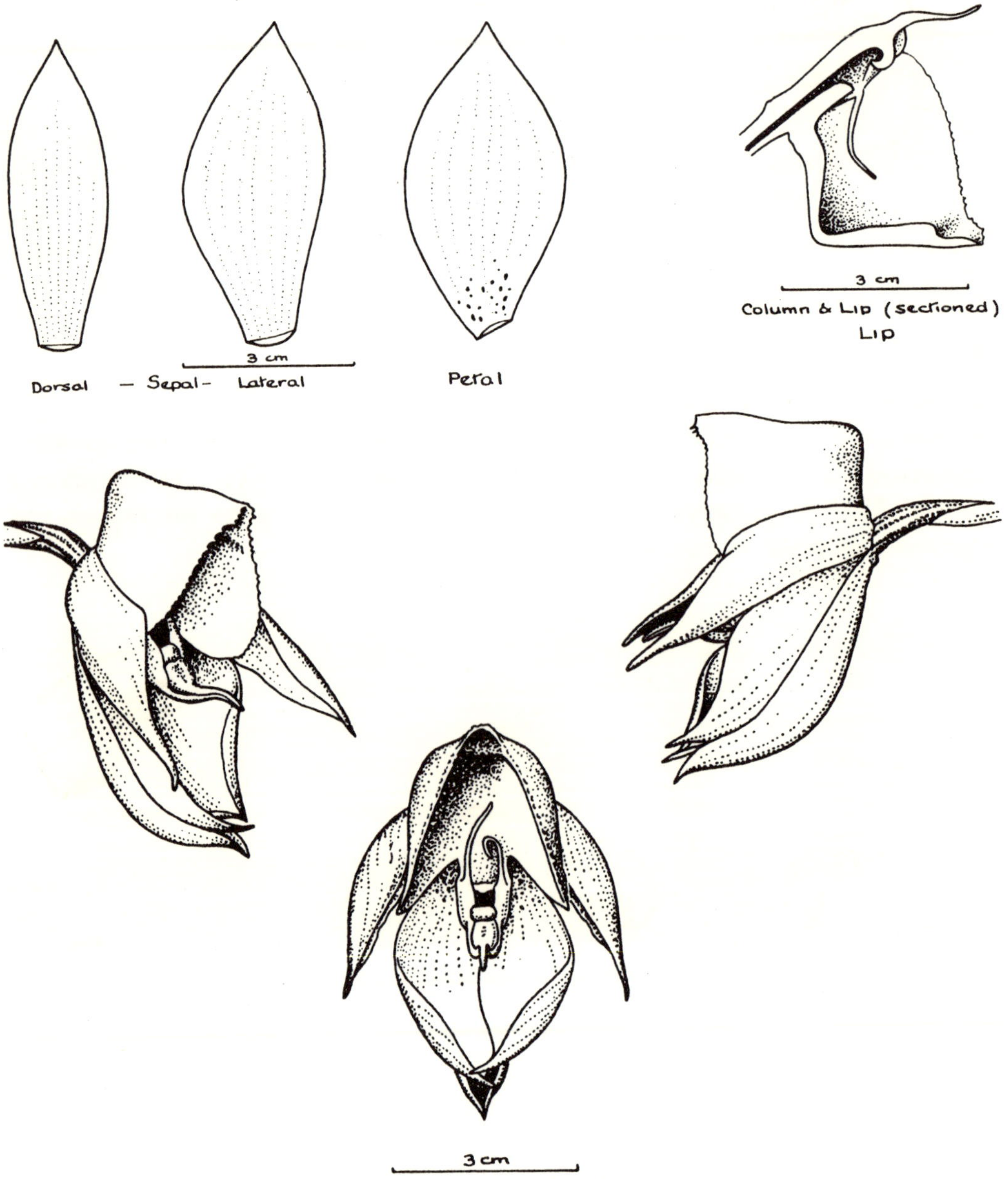

CATASETUM SPLENDENS

COCHLEANTHES DISCOLOR

COCHLEANTHES DISCOLOR (LINDL.) SCHULTES & GARAY

Synonyms: *Warrea discolor* Lindl.
Warscewiczella discolor (Lindl.) Rchb.f.
Zygopetalum discolor (Lindl.) Rchb.f.
Chondrorhyncha discolor (Lindl.) Rchb.f. P.H. Allen

Distribution: Venezuela, Cuba, Honduras, Costa Rica, Panama. In Venezuela this pseudobulbless species has been found at 1700m. in the state of Aragua not far from Caracas, and it is also on record from the state of Zulia in the Perijá Range.

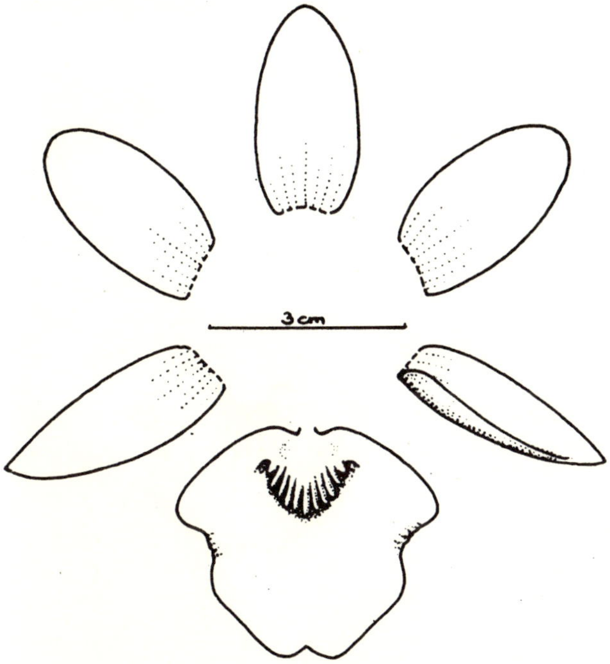

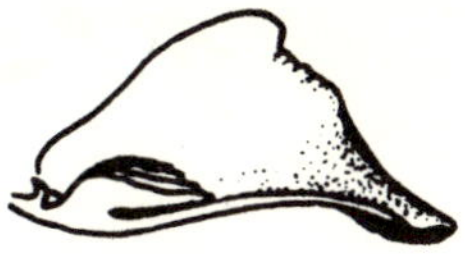

Section of Lip

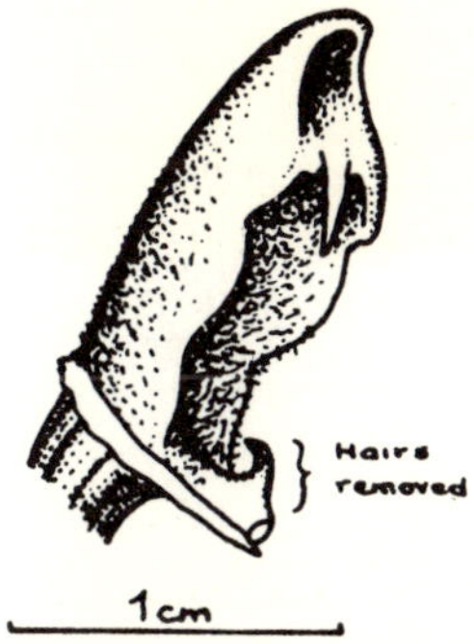

Column

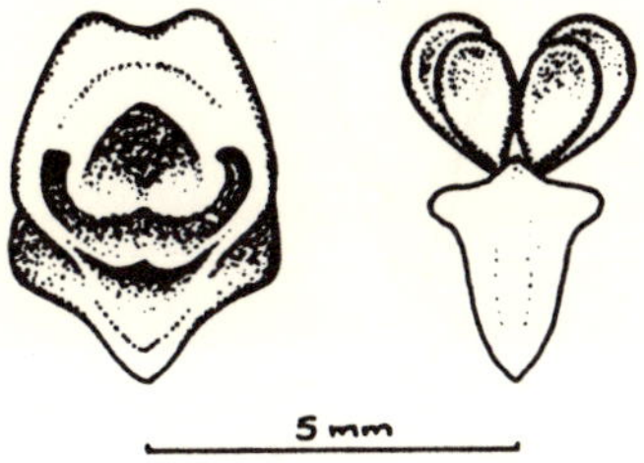

Anther & Pollinia

COCHLEANTHES MARGINATA (RCHB.F.) SCHULTES & GARAY

Synonyms: As with *Cochleanthes discolor, C. marginata* has given taxonomists much trouble in the past, with the result that *C. marginata* has had eight previous names, now treated as Synonyms, in the genera of *Warscewiczella, Huntleya*, *Warrea, Zygopetalum* and *Chondrorhyncha*.

Distribution: Venezuela, Colombia, Peru. In Venezuela it is found as a pseudobulbless epiphyte in light woods from 800 to 1500m. near Caracas and in the states of Lara and Zulia.

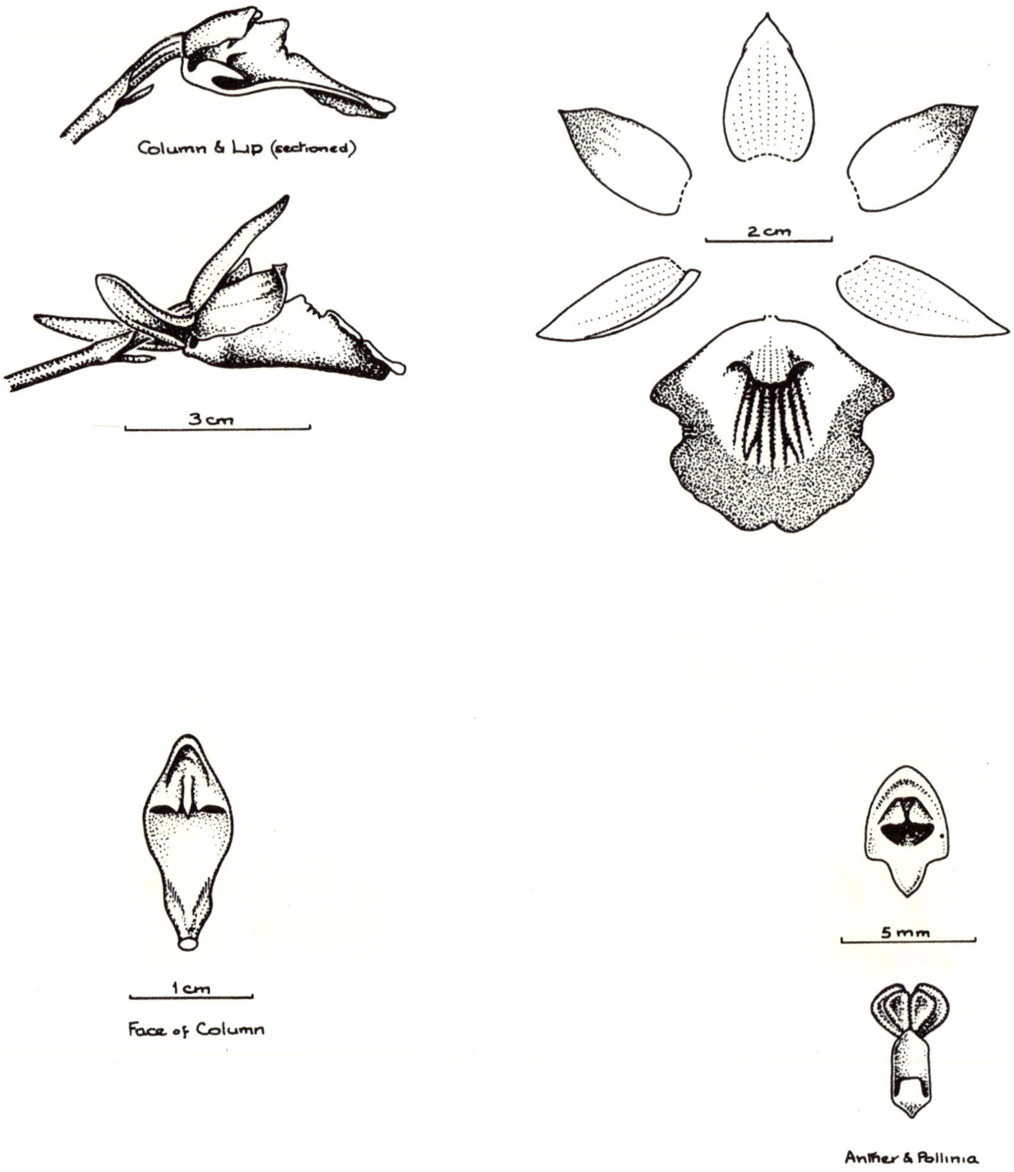

COCHLEANTHES MARGINATA

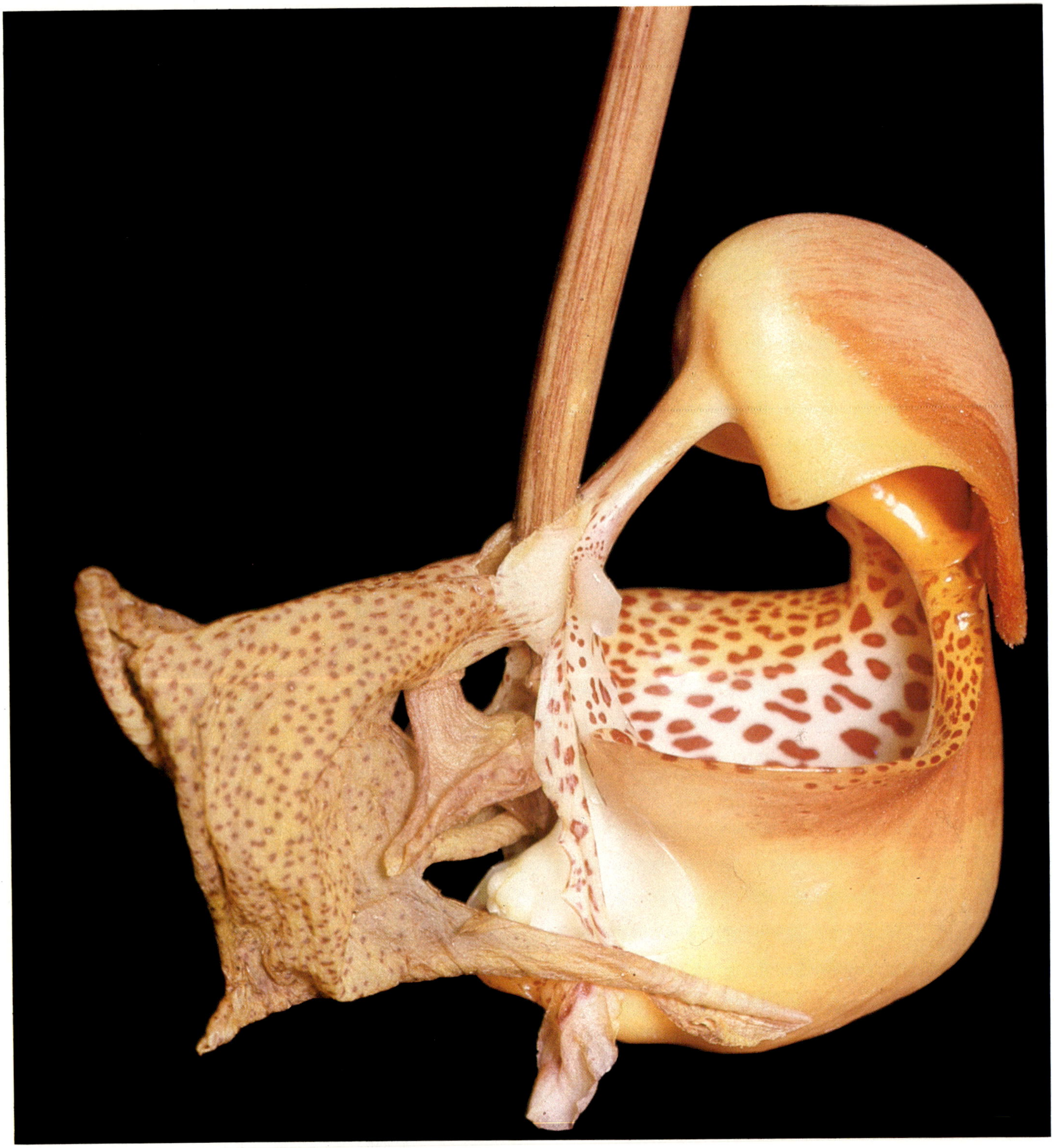

CORYANTHES BIFLORA

CORYANTHES BIFLORA BARB. RODR.

Distribution: Venezuela, Brazil, Peru. In Venezuela the habitat of this most interesting orchid seems to be limited to a small area at an altitude of about 900m. near Calderas in the state of Barinas where it grows epiphytically in light forest. The genus *Coryanthes* is famous for its most remarkable "mechanism" for attracting the special type of small bee that the species needs for its fertilization the flower, the bee falling into the suspended "bucket" that is formed by the lip of the flower and out of which it cannot escape without passing by the apex of the column of the flower where it automatically removes the pollinia and can thus transport it to the stigma of the next flower.

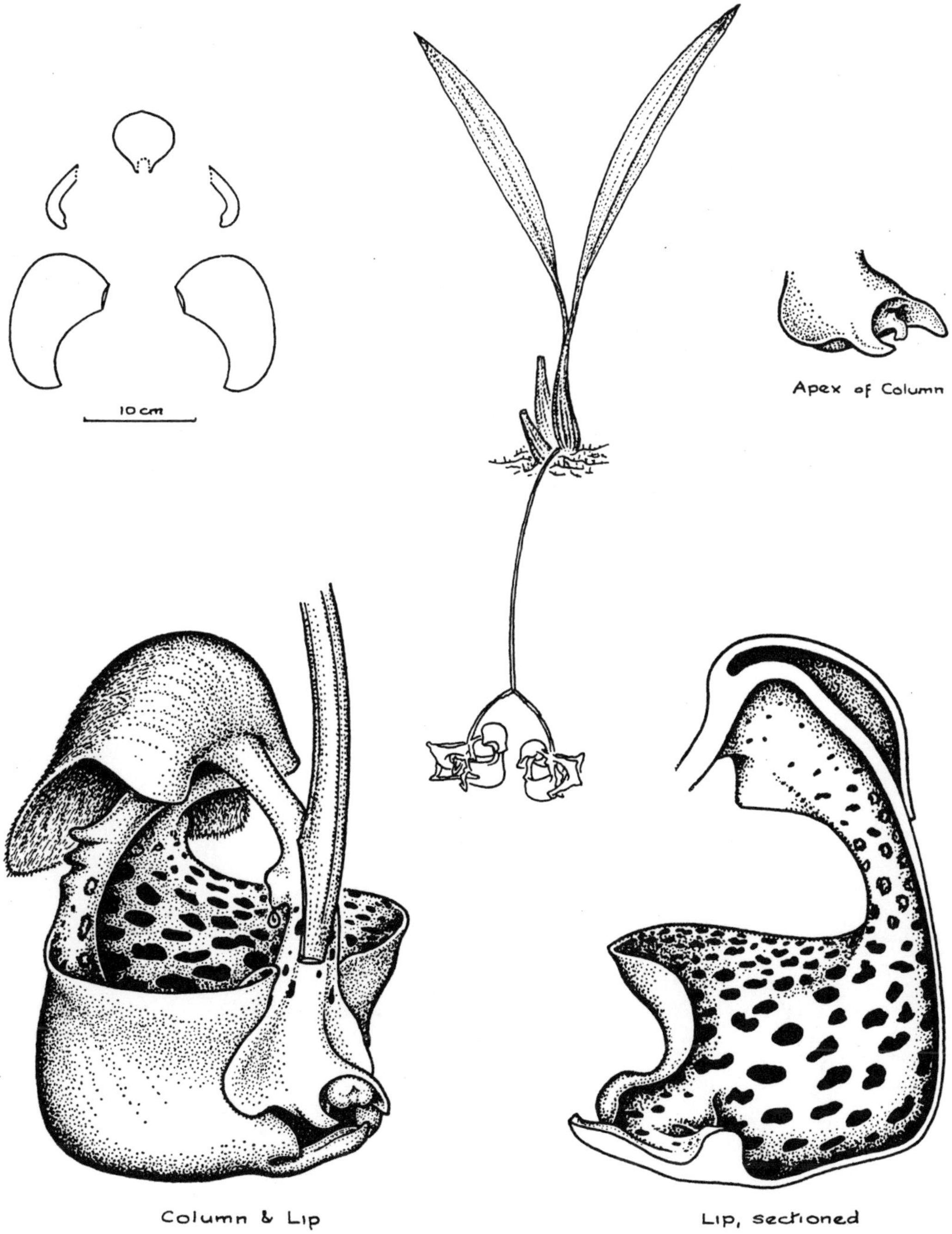

EPIDENDRUM ELONGATUM JACQ.

Synonyms: This orchid was first published under the name of *Epidendrum secundum* given it by Jacquin in 1760, and due to its great variety in color and the form of its lip went on to accumulate the extraordinary total of 35 other names in the following years, ending with *E. bulkeleyi Hawkes* of 1957, but it generally became known by its 1760 name of *Epidendrum secundum*. In 1974, however, Garay, in his book on the Flora of the Lesser Antilles pointed out that this name had already been given to a quite different species and therefore another Jacquin name, *Epidendrum elongatum*, of 1789, is the name used here.

Distribution: Common and widespread throughout the whole American tropics. In Venezuela it grows mainly as a terrestrial, but is not uncommon also as an epiphyte, and there are few parts of Venezuela between the altitudes of 700 and 2000 m. where it is not present.

As mentioned above, there are many color varieties or "forms" of this species, ranging from white to yellow, orange, pink, red or purple, with blue the only color not reported. The yellow form shown here comes mainly from the Venezuelan Andes.

Column & Lip.

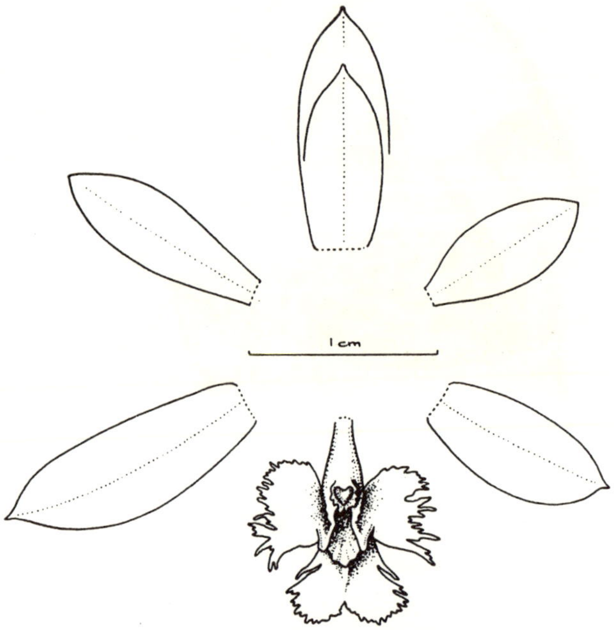

EPIDENDRUM ELONGATUM

GALEANDRA MINAX

GALEANDRA MINAX RCHB.F.

Distribution: Venezuela, Brazil. In Venezuela this species seems to be rare, and apart from a vague mention of it by the 19th century botanist A. Cogniaux, with the exact locality not mentioned, it seems to have been recorded only from Cerro Duida (Federal Territory of Amazonas) where it was found in 1975, at about 1200m, by botanist S.S. Tillett. It was growing in a hole in the trunk of an old tree. The slender erect secondary stems were 30cm. tall and the terminal inflorescence extended a further 10cm. in the form of a short-branched panicle of 3 flowers.

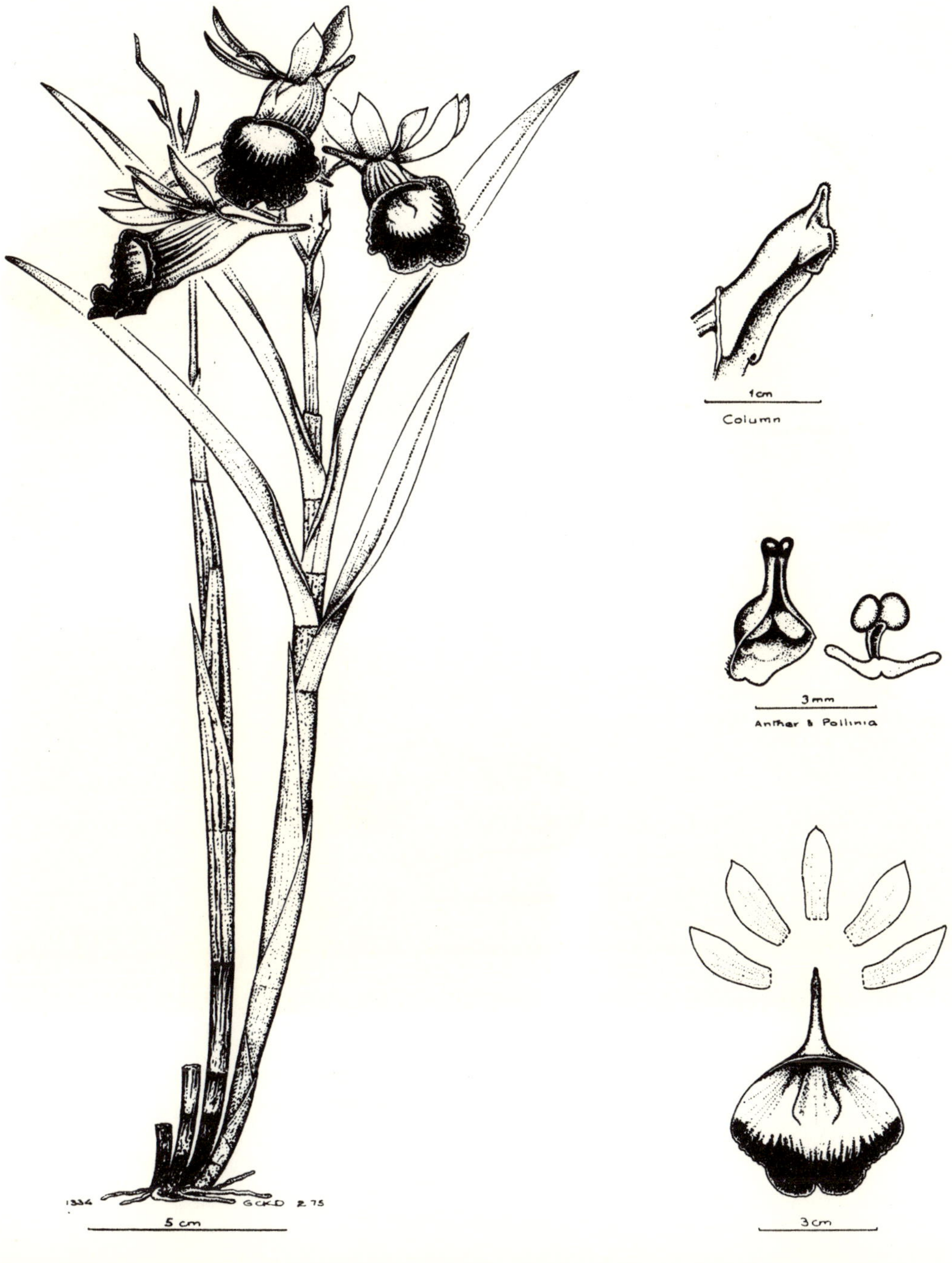

MASDEVALLIA TRIANGULARIS LINDL.

Distribution: Venezuela and Colombia. In Venezuela this species is known from the Coastal Range area of the states of Aragua and Carabobo at altitudes from 1700 to 2300m., and in the Andes in the states of Lara, Mérida and Táchira from 1800 to 3000m.

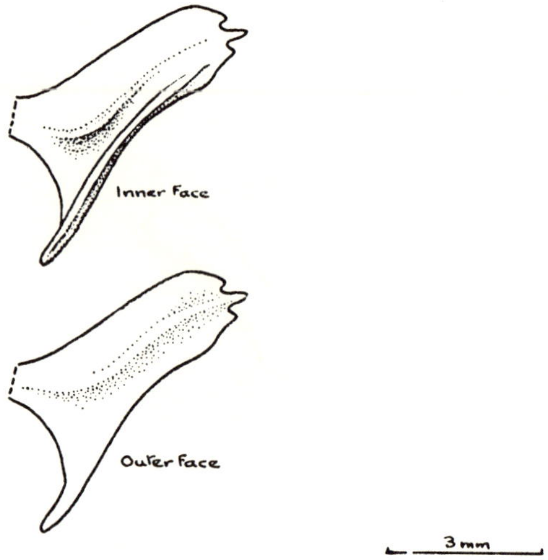

Petal

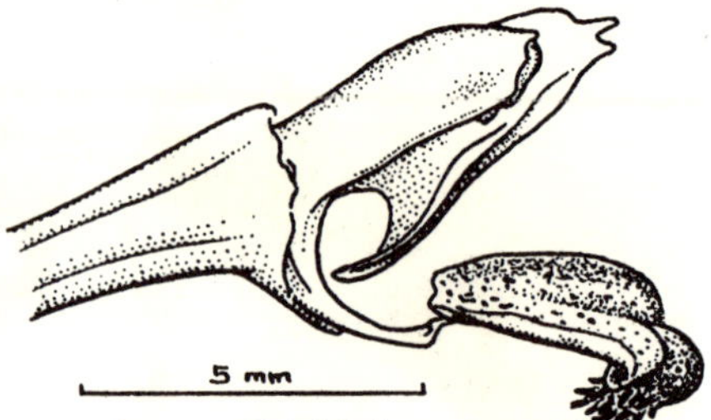

Column, Petal & Lip

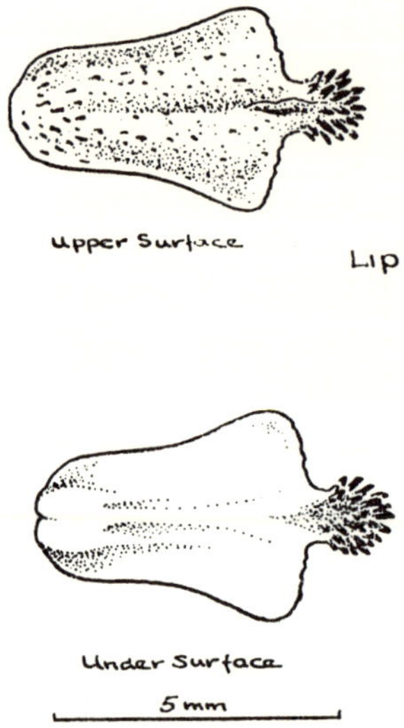

Lip

MASDEVALLIA TRIANGULARIS

MAXILLARIA DESVAUXIANA

MAXILLARIA DESVAUXIANA RCHB. F.

Synonyms: *Maxillaria coriacea* Barb. Rodr,
Maxillaria petiolaris A. Rich.

Distribution: Venezuela, Guyana, Surinam, Brazil. In Venezuela epiphytic (and in exposed dry places also lithophytic) at various altitudes from 150 to 2000m. in the state of Miranda and the Federal Territory of Amazonas on Cerros Autana, Sipapo and Duida.

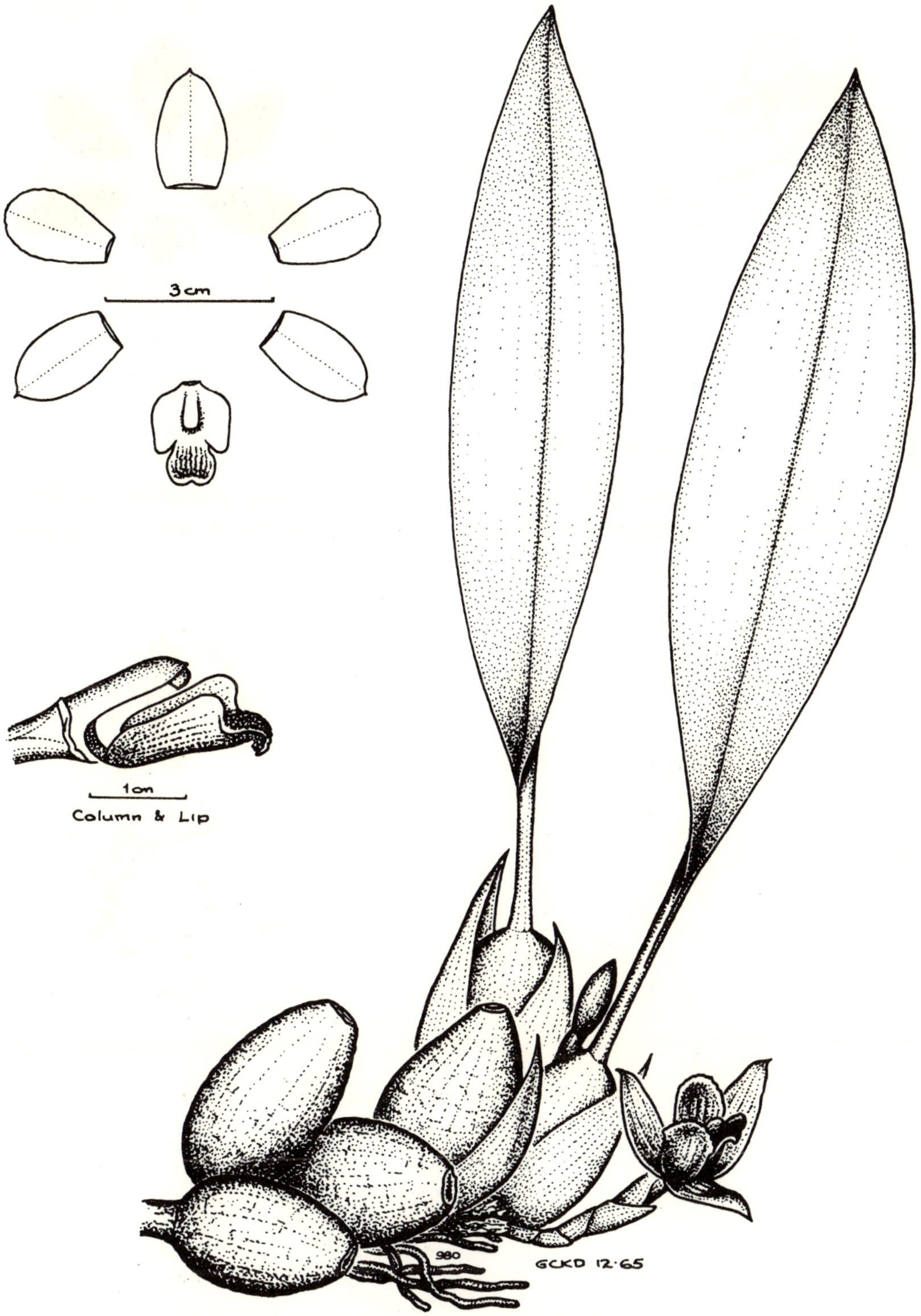

ONCIDIUM LANCEANUM LINDL.

Synonym: *Lophiaris fragrans* Raf.
Distribution: Venezuela, Trinidad, Guyana, Surinam, Brazil. In Venezuela this species grows both terrestrially and epiphytically at altitudes from 50 to 200m. in the states of Monagas, Sucre and Bolívar, and in the Federal Territories of Amazonas and Delta Amacuro.

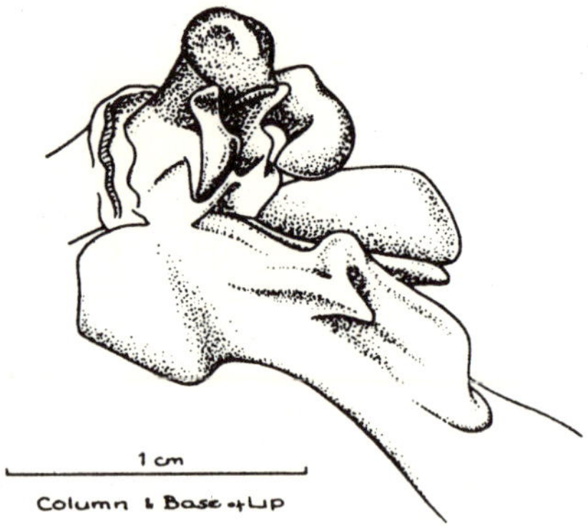

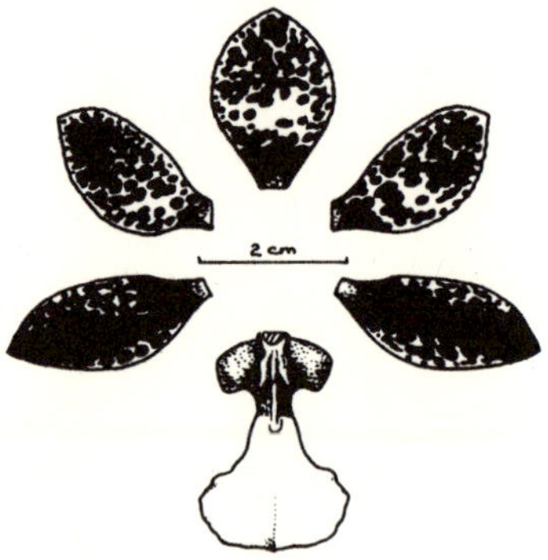

ONCIDIUM MIMETICUM STACY

Distribution: Venezuela, Colombia, Ecuador. In Venezuela this orchid has been found growing as an epiphyte at the headwaters of the Quinimarí River, at an altitude of 2450m, in the state of Táchira.

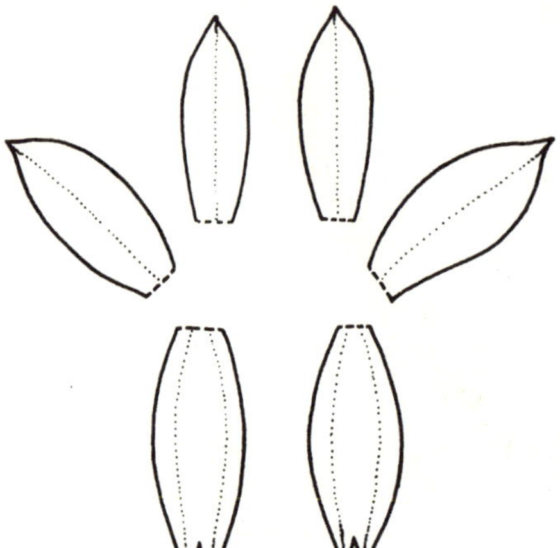

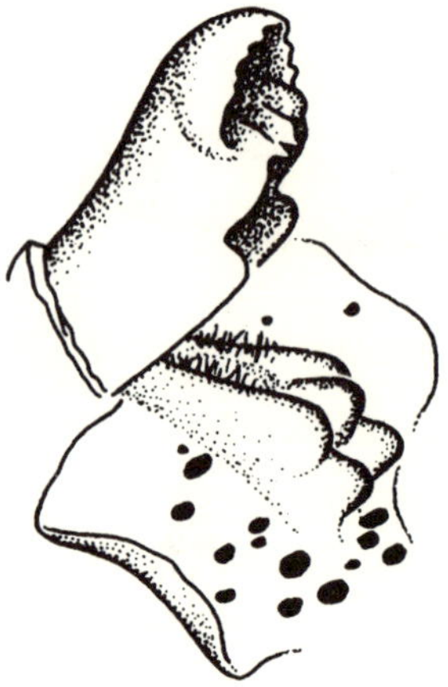

Column & Base of Lip (depressed)

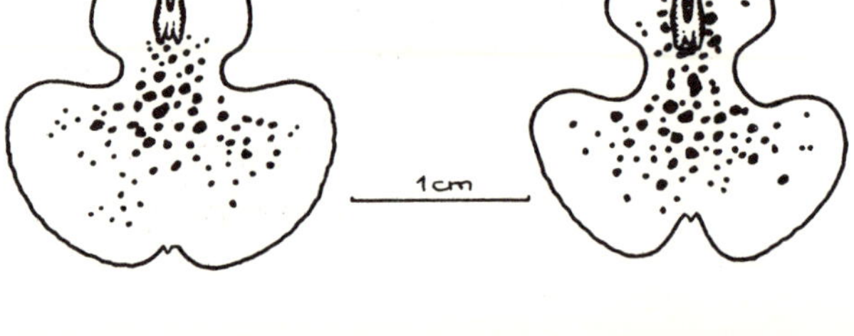

ONCIDIUM LANCEANUM

ONCIDIUM MIMETICUM

POLYCYCNIS MUSCIFERA

POLYCYCNIS VITTATA

POLYCYCNIS MUSCIFERA (LINDL.) RCHB. F.

Synonyms: *Cycnoches muscifera* Lindl. & Paxt.
Cycnoches buchtienii Krzl.

Distribution: Venezuela, Panama, Colombia, Ecuador, Peru, Bolivia. In Venezuela this species has been found growing terrestrially in the steep bank of a ravine at an altitude of 590m. near San Cristóbal in the state of Táchira. The inflorescences are 50 to 70 cm. long, the basal peduncle erect, the rachis variably arching or pendent. The plicate leaves can be up to 1m. tall.

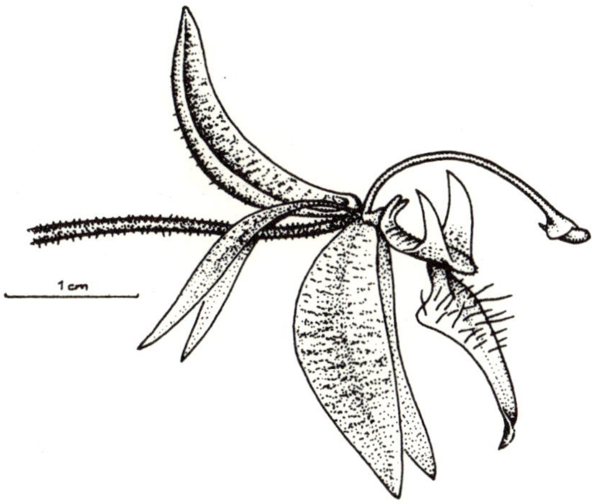

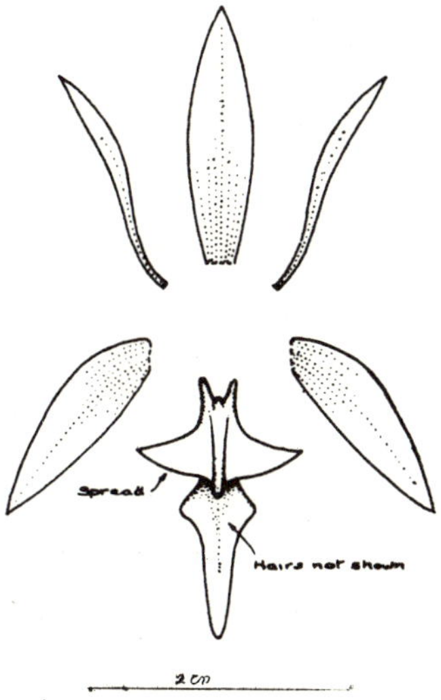

POLYCYCNIS VITTATA (LINDL.) RCHB. F.

Synonym: *Houlletia vittata* Lindl.

Distribution: Venezuela, Guyana, Colombia, Peru, Brazil. In Venezuela this species occurs as an epiphyte in the state of Bolívar at altitudes from 100 to 350m., and in the Federal Territory of Amazonas at 100-200m. In contrast to *Polycycnis muscifera*, the many-flowered inflorescence of *Polycycnis vittata* is stoutly erect, and the exposed, branching and spiky roots at the base of the pseudobulbs are a prominent character.

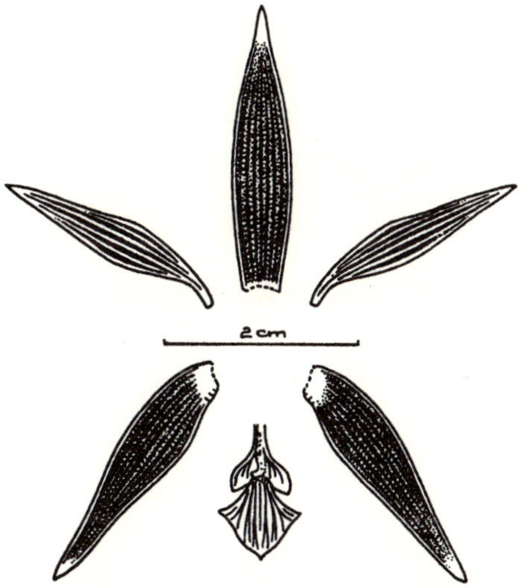

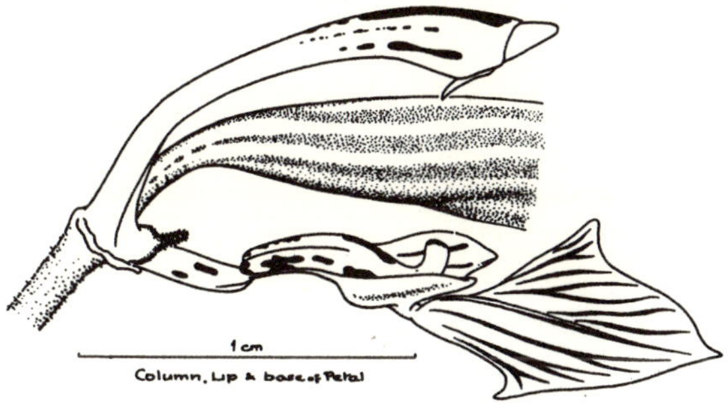

MAXILLARIA GUAREIMENSIS RCHB.F.

Synonym: *Maxillaria spilotantha* Rchb.f.
Distribution: Venezuela, Colombia, Ecuador. In Venezuela this species occurs as an epiphyte from 600 to 1500m. altitude in the coastal parts of the states of Aragua and Falcón, and at an altitude of 700m. in the state Bolívar on Cerro Guaiquinima.

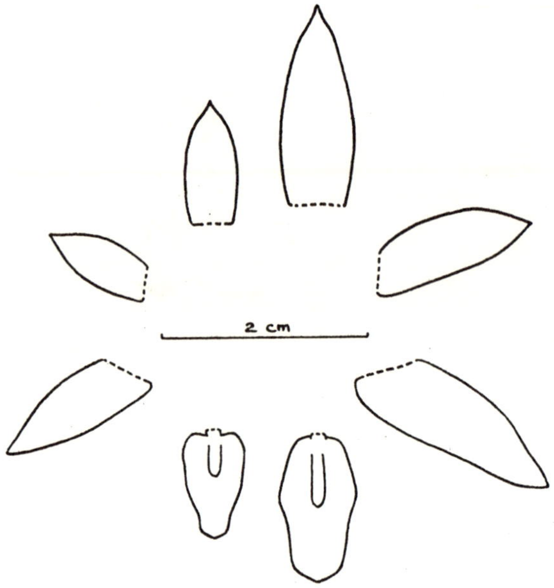

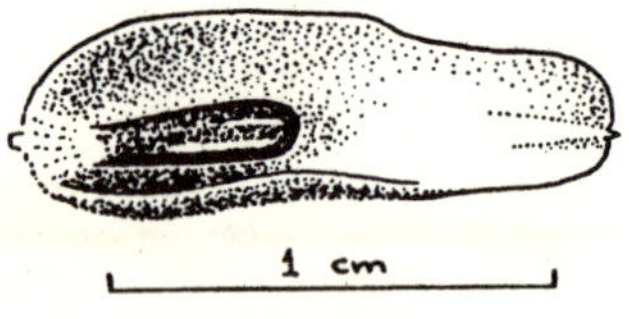

Lip

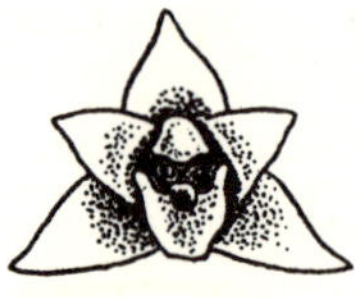

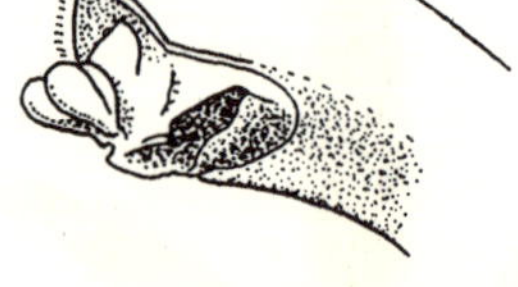

Tip of Column
Anther removed, showing Pollinia

Column & Lip

MAXILLARIA GUAREIMENSIS

MAXILLARIA HISTRIONICA

MAXILLARIA HISTRIONICA (RCHB. F.) L.O. WMS.

Synonyms: *Ornithidium histrionicum* Rchb.f.
Maxillaria aristeguietae Foldats

Distribution: Venezuela and Mexico. In Venezuela this epiphytic species has been found in the Coastal Range, at altitudes between 1500m. and 2000m., in the states of Aragua and Carabobo.

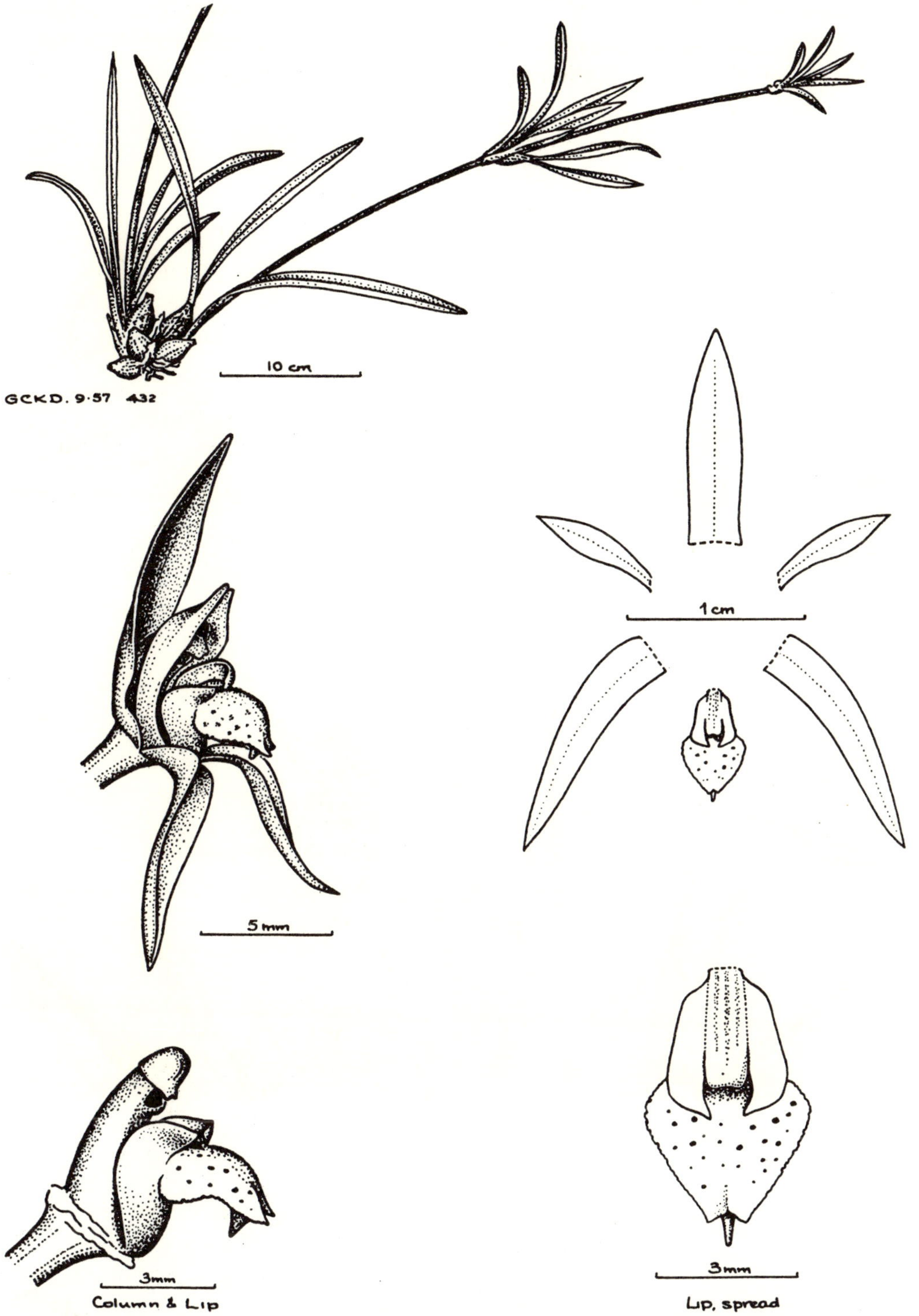

MAXILLARIA PARKERI HOOK.

Synonyms: *Menadena parkeri* Raf.
Maxillaria hirtilabia Lindl.
Maxillaria lorifolia Rchb.f.
Maxillaria kegelii Rchb.f.
Maxillaria loretoensis C. Schweinf.

Distribution: Venezuela, Guyana, Surinam, Peru, Brazil. In Venezuela it occurs as an epiphyte at altitudes from 200 to 1200m. in the state of Bolívar in the Gran Sabana and on Cerro Guaiquinima, Auyantepui, Kukenan etc. and in the Federal Territory of Amazonas can be found at 100 to 200m. on the Upper Orinoco.

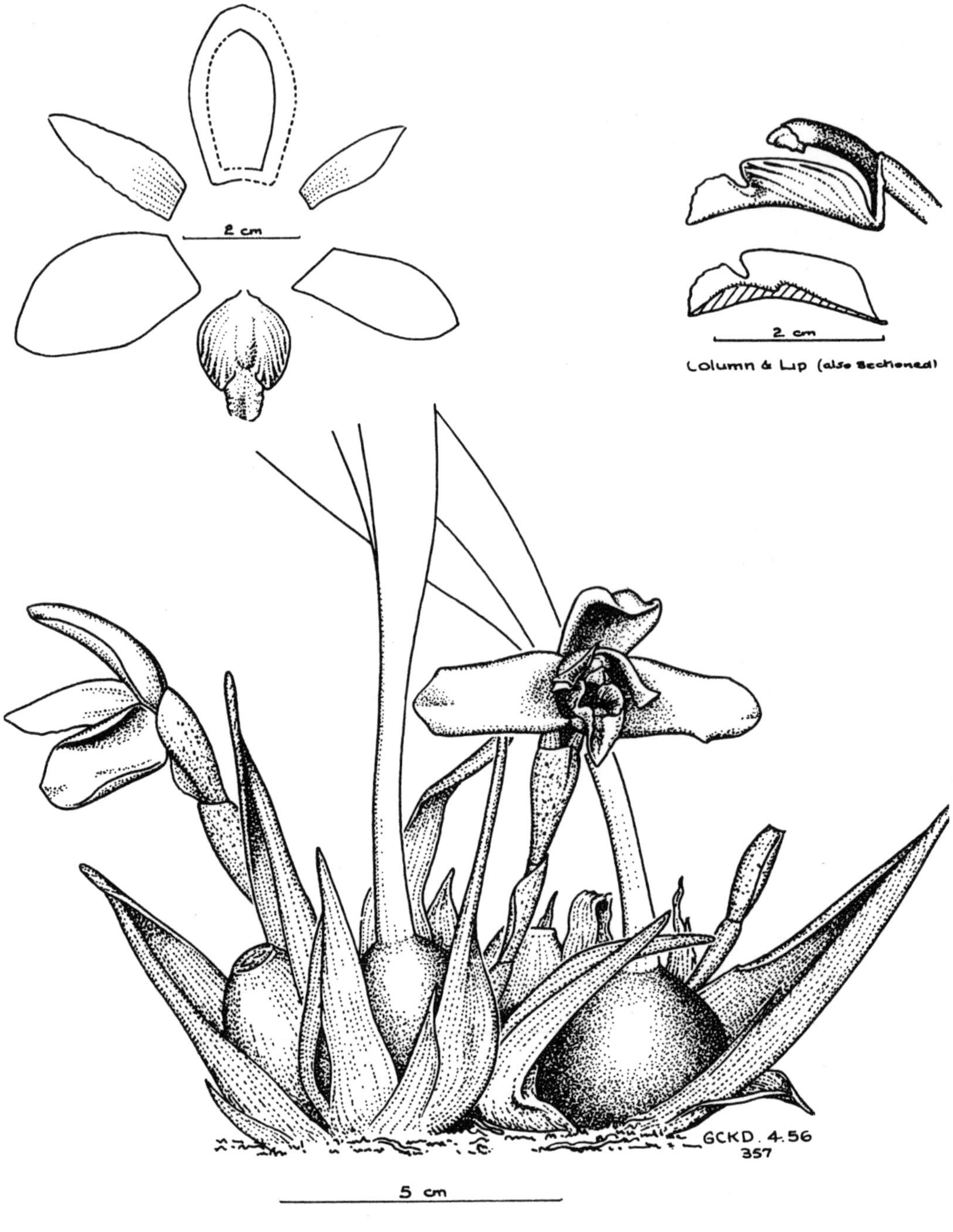

MAXILLARIA PARKERI

MAXILLARIA SETIGERA

MAXILLARIA SETIGERA LINDL.

Synonyms: *Maxillaria leptosepala* Hook.
Maxillaria caloglossa Rchb.f.

Distribution: Venezuela, Guyana, Colombia. In Venezuela it grows as an epiphyte from altitudes of 250 to 600m., having been recorded from the state of Táchira in the Andes, from the state of Bolívar and from the Federal Territory of Amazonas.

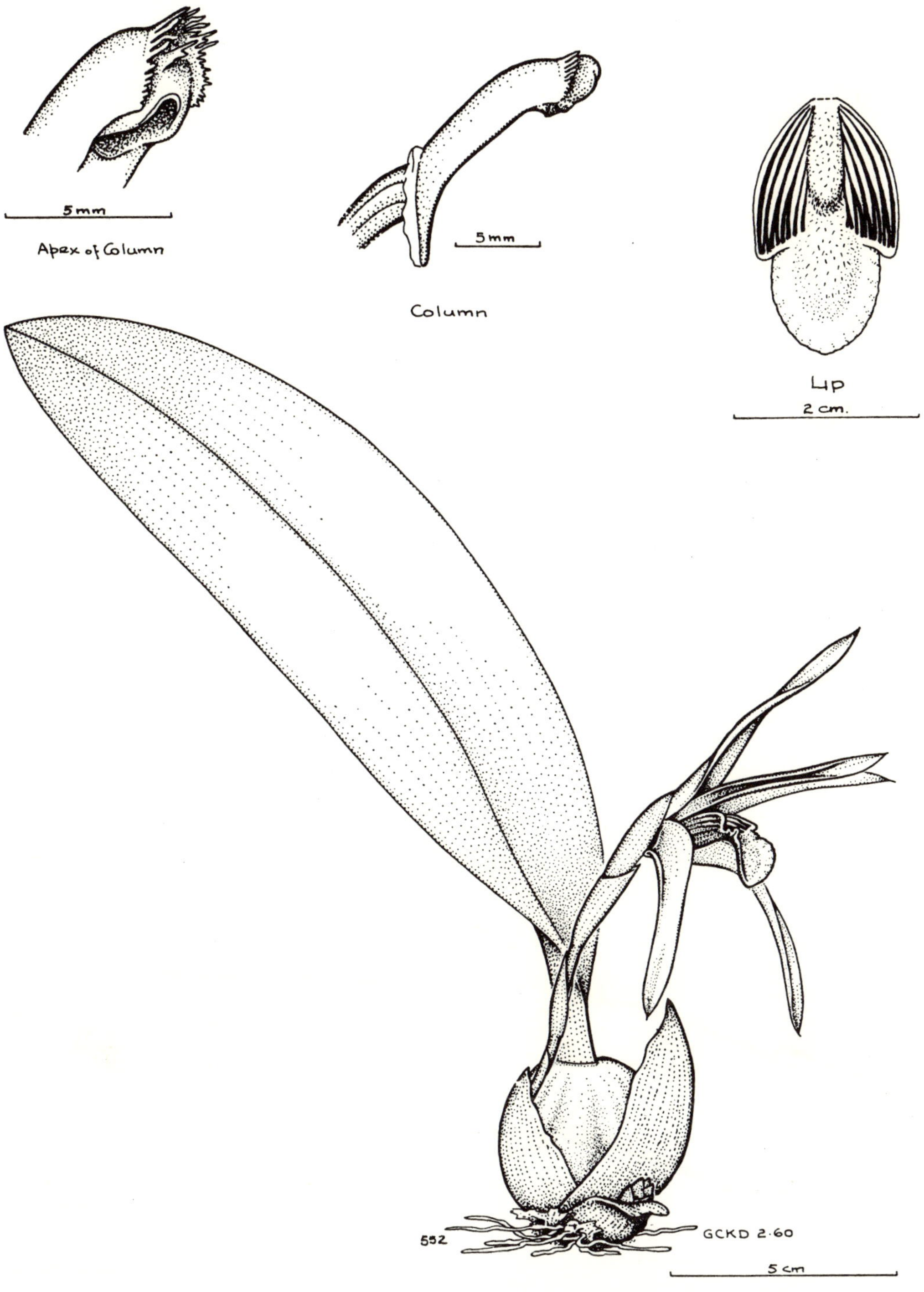

MAXILLARIA SPLENDENS POEPP. & ENDL.

Synonym: *Maxillaria buchtienii* Schltr.
Distribution: Venezuela, Surinam, Colombia, Ecuador, Peru, Brazil. In Venezuela this species has a vertical range from an altitude of 400m. to an altitude of 1400m., and while it grows mainly as an epiphyte it can also be found occasionally growing as a terrestrial. In the state of Bolívar it has been found near the base of Cerro Sarisariñama and in the Gran Sabana near the Brazilian border; and in the Federal Territory of Amazonas it has been recorded from the summit of Cerro Autana and Sierra Parima. The flowers of this species are variable in size, with sepals up to 14cm. long.

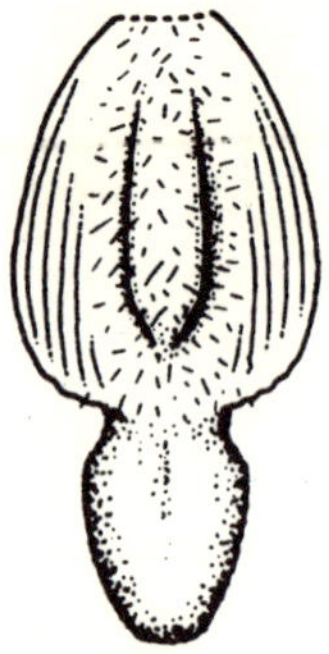

1 cm

Lip

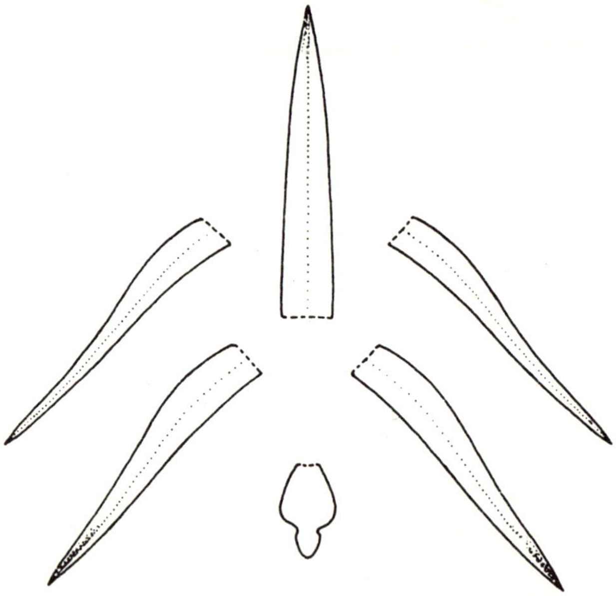

3 cm

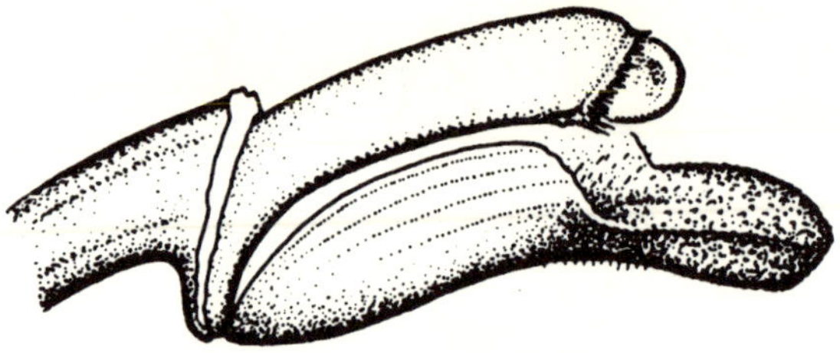

1 cm

Column & Lip

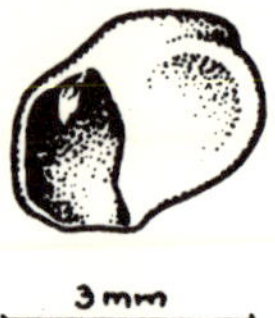

3 mm

Anther & Pollinia

MAXILLARIA SPLENDENS

MAXILLARIA TRILORIS

MAXILLARIA TRILORIS E. MORR.

Distribution: Endemic to Venezuela where it occurs as an epiphyte in the state of Miranda (Guatopo National Park) at around 700m., and in the state of Bolívar at altitudes of 1500 to 2000m.

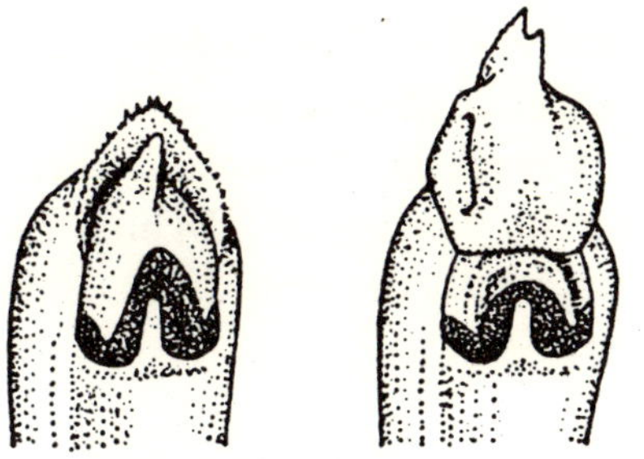

Apex of Column, without & with Anther

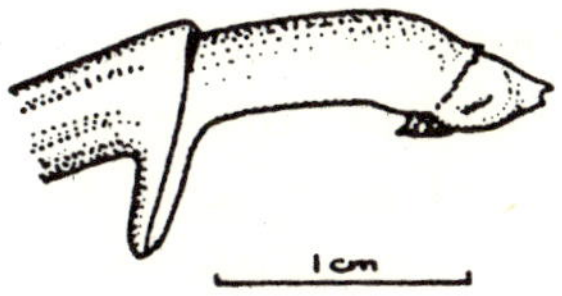

Column

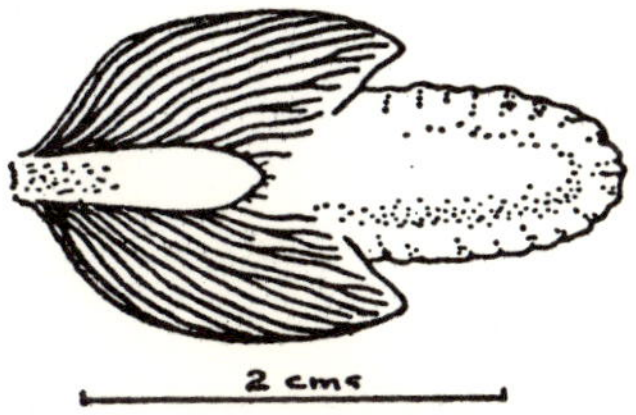

Lip, spread

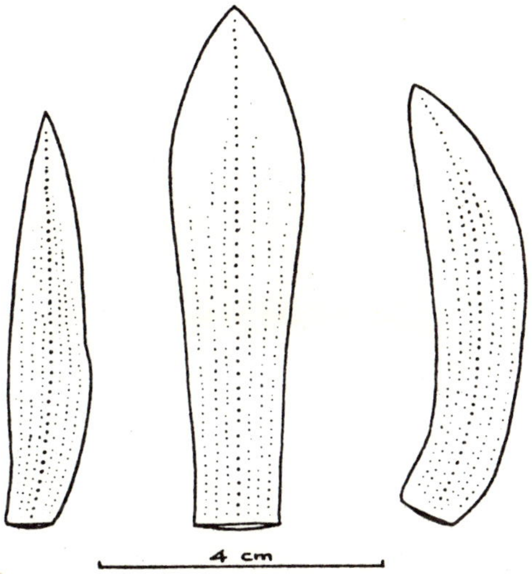

Petal Dorsal Sepal Lateral Sepal

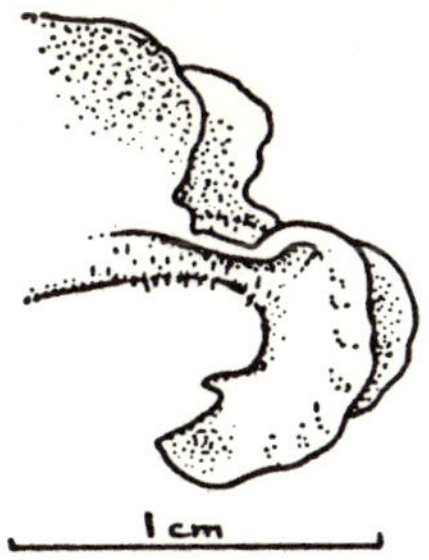

Apex of Lip

PHRAGMIPEDIUM CAUDATUM LINDL.

Synonyms: Early confusion in the classification of orchids with "slipper lips" resulted in this species having been recorded at various times under the generic names of *Cypripedium, Selenipedium, Phragmipedium, Paphiopedilum* and *Phragmopedilum*. There is also a lipless form that is known to grow intermixed with the normal form and which is considered by many to be merely a variety of *Phragmipedium caudatum*. This form was also known as *Uropedium lindeni*, but has recently been named *Phragmipedium lindeni* by Dressler & Williams.
Distribution: Guatemala, Costa Rica, Panamá, Venezuela, Colombia, Ecuador and Perú. In Venezuela *P. caudatum* is mentioned in Venezuelan Orchids Illustrated as being a native of the state of Mérida at 5,600 ft. altitude (1,700m.) and is similarly noted in the Flora de Venezuela of 1969.

Typical leaf Apices

5 m.m.

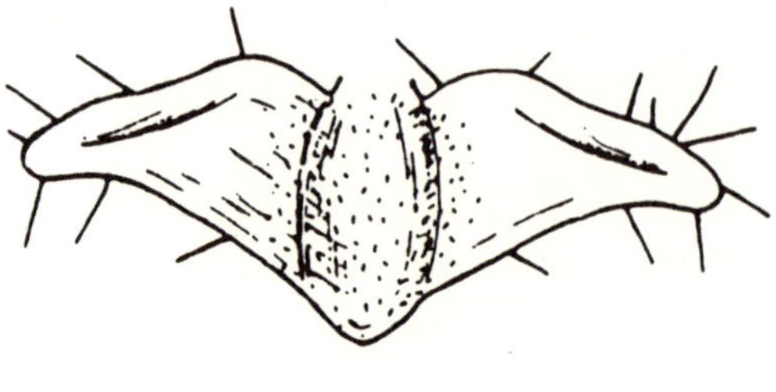

3) Staminode

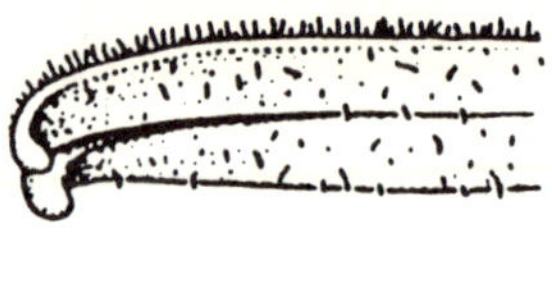

Apex of Petal

3 m.m.

PHRAGMIPEDIUM CAUDATUM

ONCIDIUM FALCIPETALUM

ONCIDIUM FALCIPETALUM LINDL.

Synonyms: *Oncidium ionodon* Rchb.f.
Oncidium davisii Rchb.f.
Cyrtochilum falcipetalum Krzl.

Distribution: Venezuela, Colombia, Ecuador, Peru. In Venezuela this species grows epiphytically near Caracas, from 1000m. to 2300m. altitude and from 1500 to 2900m. in the states of Aragua, Mérida, Táchira and Trujillo.

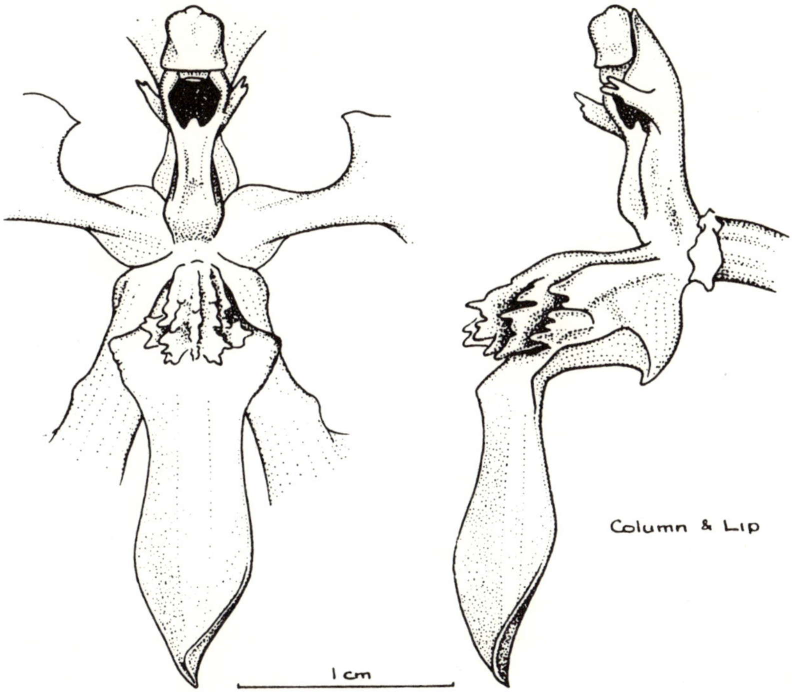

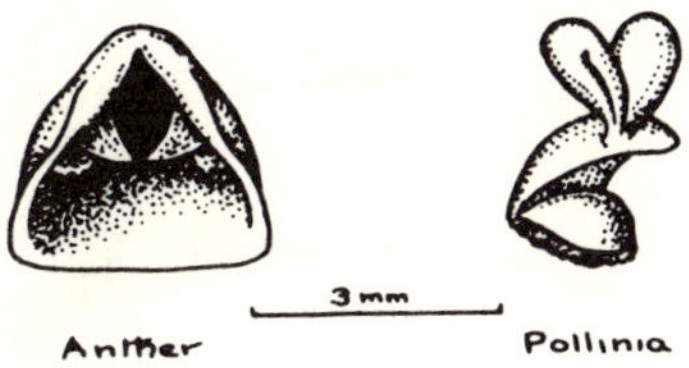

RODRIGUEZIA LEEANA RCHB. F.

Synonyms: *Burlingtonia leeana* Williams
Burlingtonia negrensis Barb. Rodr.
Rodriguezia negrensis Cogn.

Distribution: Venezuela, Brazil. In Venezuela it has been found as an epiphyte at low altitudes in the Federal Territory of Amazonas, near the Mavaca and Padamo Rivers.

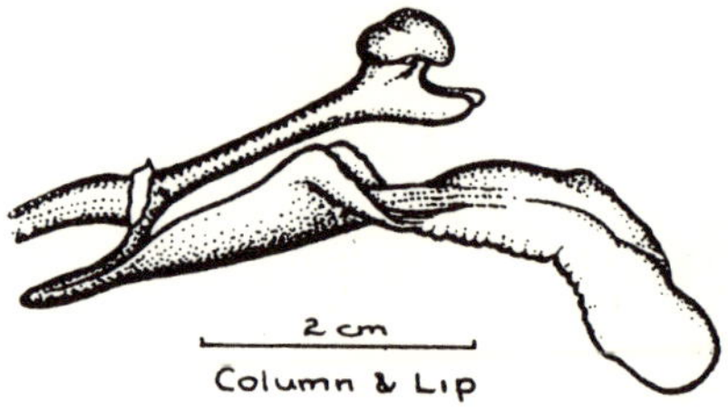

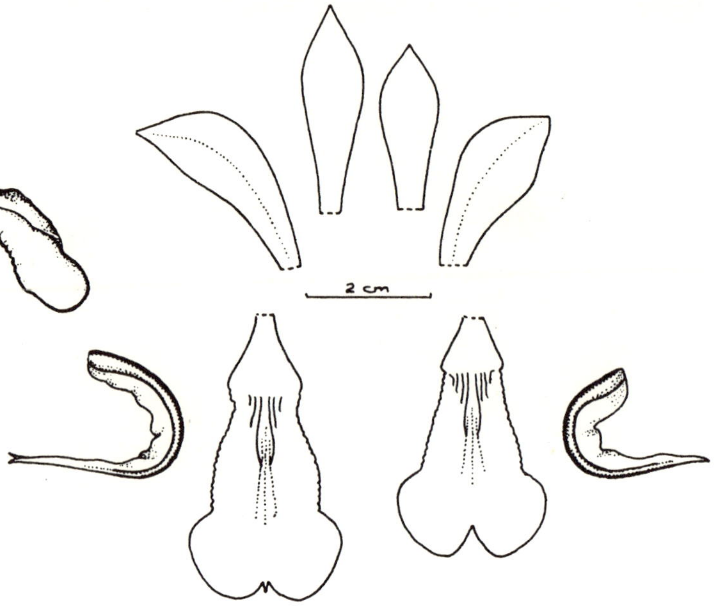

SARCOGLOTTIS GRANDIFLORA (LINDL.) KL.

Distribution: Fairly common throughout the American tropics. In Venezuela this species occurs as a tall and handsome terrestrial that can be found at almost sea-level on the coast not far from Caracas, but it also grows at much higher altitudes, such as at 1300m. on the flank of the Avila Range also near Caracas. It has also been found at about 300m. altitude in the state of Bolívar on the Paragua River, 800m. altitude on the flank of Cerro Venamo, and at about 200m. on the Caura River. For long this species erroneously bore the name of *Spiranthes acaulis*, but this has now been shown by Garay to be the name of a different orchid.

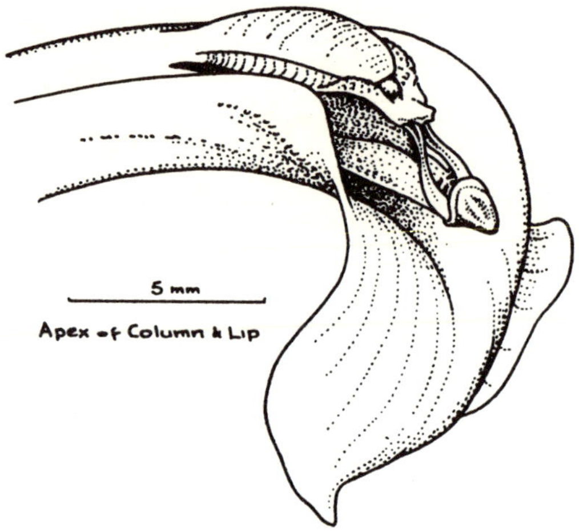

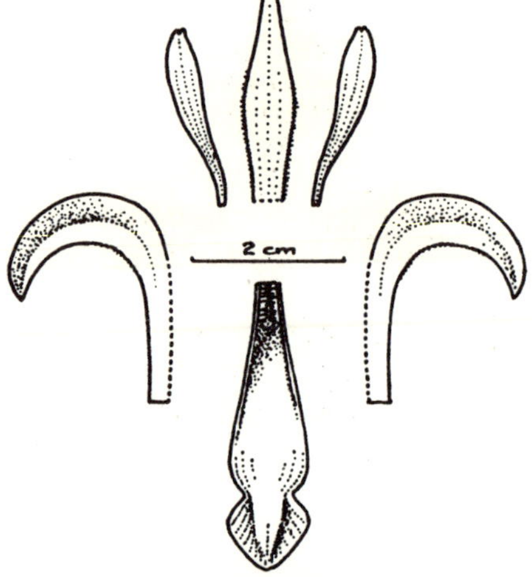

RODRIGUEZIA LEEANA

SARCOGLOTTIS GRANDIFLORA

STANHOPEA WARDII

STENIA PALLIDA

STANHOPEA WARDII LODD. EX LINDL.

Synonyms: *Stanhopea aurea* Lodd.
Stanhopea venusta Lindl.
Stanhopea amoena Kl.

Distribution: Venezuela to Peru, Mexico to Panama. In Venezuela it grows as an epiphyte at altitudes of 500 to 1800m. near Caracas and in the Coastal Range states of Aragua, Carabobo and Miranda, and also at about 1000m. altitude in the Perijá Range of the state of Zulia.

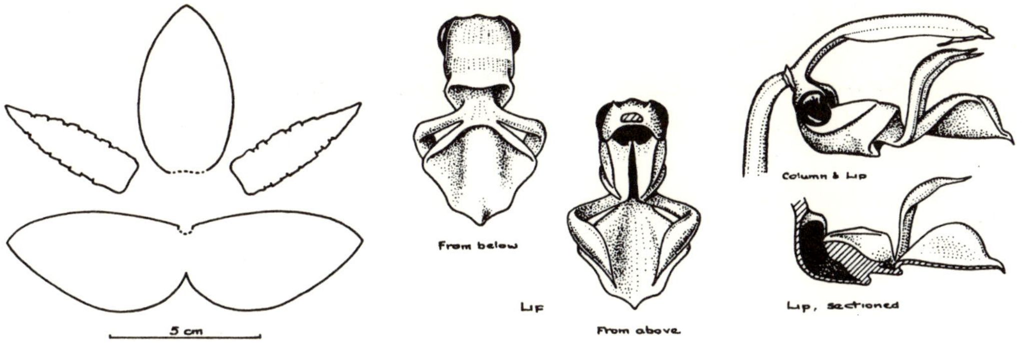

STENIA PALLIDA LINDL.

Distribution: Venezuela, Guyana, Trinidad, Peru, Brazil. In Venezuela it occurs as a widespread species with no pseudobulbs, at very variable altitudes, having been recorded from an altitude of 2000m. above Caracas and at about 600 to 1400m. in the Coastal Range states of Miranda, Aragua and Carabobo and in the Andes of the state of Lara at 2200m. It has also been recorded from 300m. altitude in the Península of Paria in the state of Sucre, and at about 400m. in the Altiplanicie de Nuria and the Río Caura in the state of Bolívar.

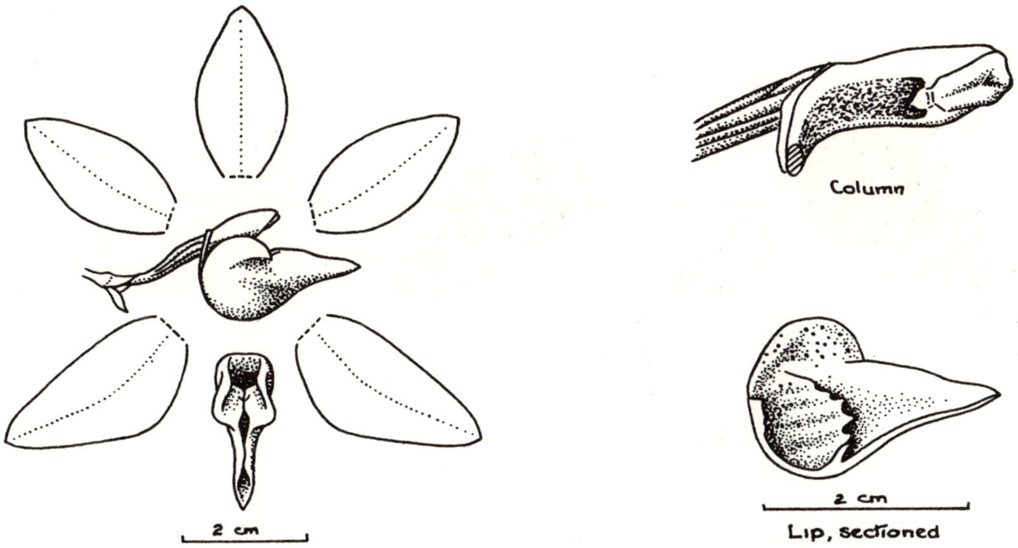

ONCIDIUM PAPILIO LINDL.

Synonym: *Psychopsis picta* Raf.

Distribution: Venezuela, Trinidad, Colombia, Ecuador, Peru. In Venezuela it can be found as an epiphyte at altitudes from 800 to 1200m. near Caracas, in the coastal states of Carabobo and Miranda and in the Andean states of Lara, Trujillo and Mérida.

This very striking species is aptly named because its flowers, borne on long and flexible stems, seem to float in the air like brightly colored butterflies. Their floral make-up is also very strange as they seem to have three very narrow sepals like antennae and two quite wide petals, but careful checking shows that two of the three antennae are petals and the third is actually the dorsal sepal, while the apparent petals are the lateral sepals! The normal flower measures about 12 to 15 cm. in spread, but exceptionally fine flowers almost twice as big have been seen.

This species is often mistaken for the very similar *Oncidium kramerianum*, but can be easily distinguished from this because the upper part of the stem of *O. papilio* is strongly flattened whereas that of *O. Kramerianum* is round.

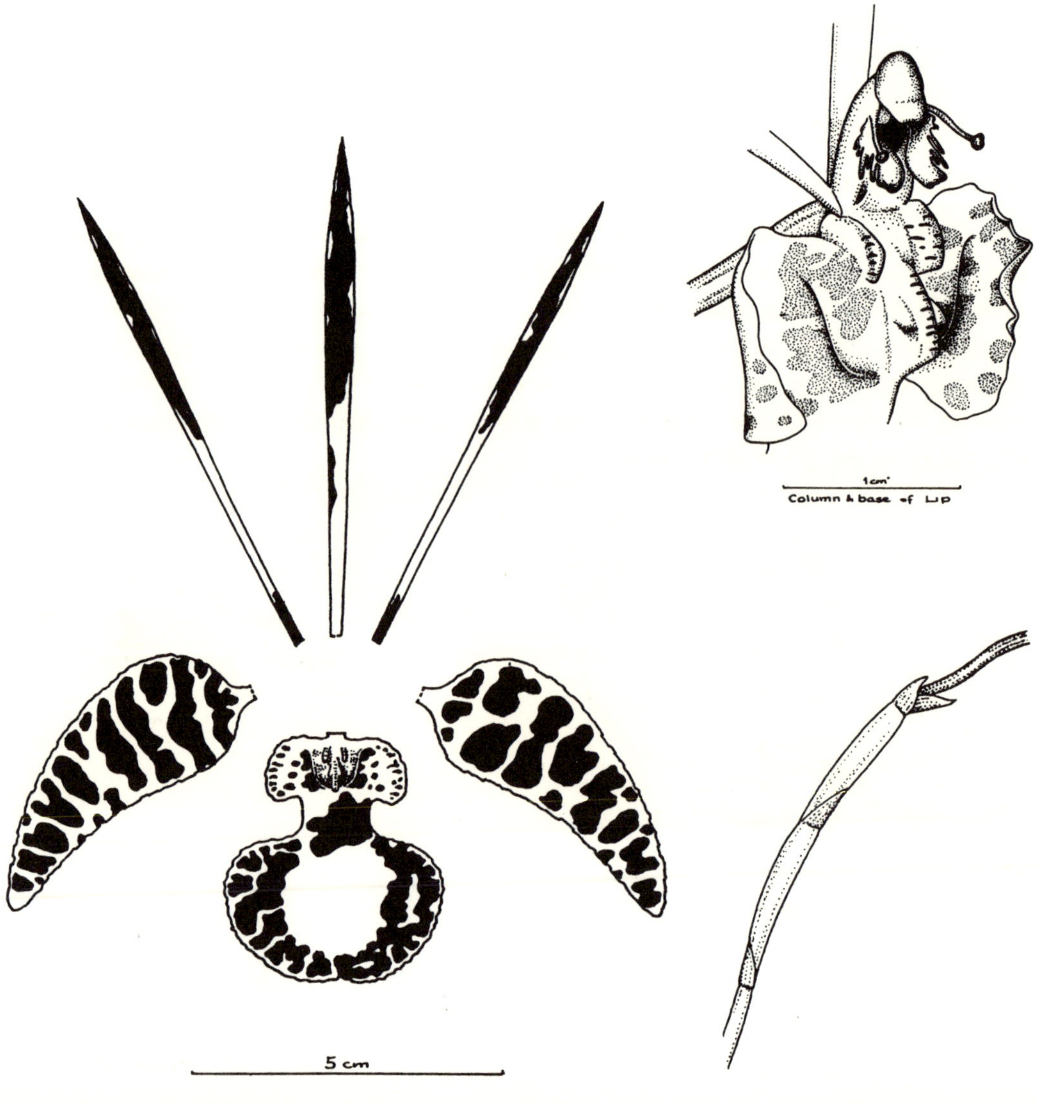

ONCIDIUM PAPILIO

PAPHINIA LINDENIANA

PAPHINIA LINDENIANA RCHB.F.

Synonym: *Lycaste lindeniana* (Rchb.f.) Nichols

Distribution: Venezuela, Guyana, Brazil. In Venezuela this epiphyte is decidedly rare, having been recorded from only three sites, all in the Federal Territory of Amazonas: one at 100m. altitude on Cerro Yapacana, another at 150m. altitude at the upper end of the Lower Siapa River, and a third at 600m. altitude at the top end of the Putaco River close to the Brazilian border. With its pendent raceme of fairly large dark red flowers streaked with white, it is certainly a handsome and striking species, even if it might not be called a truly beautiful one.

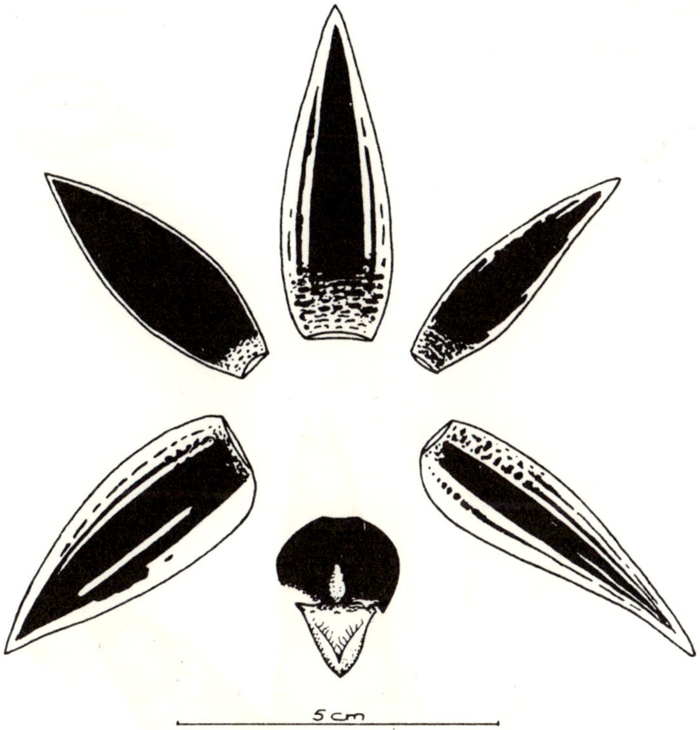

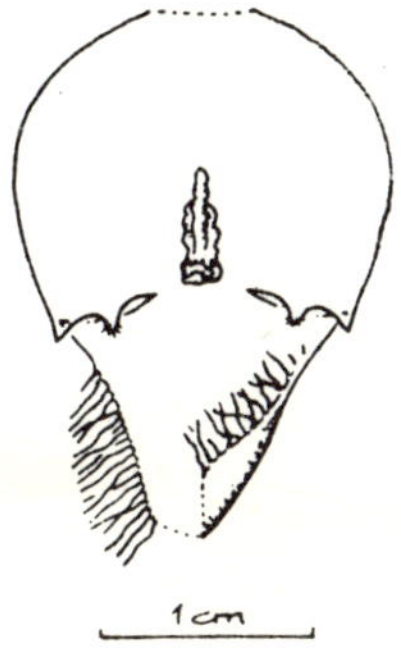

Lip

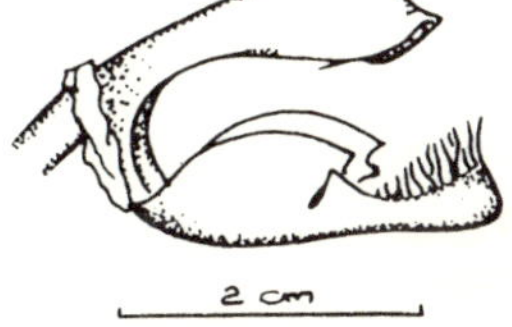

Column & Lip

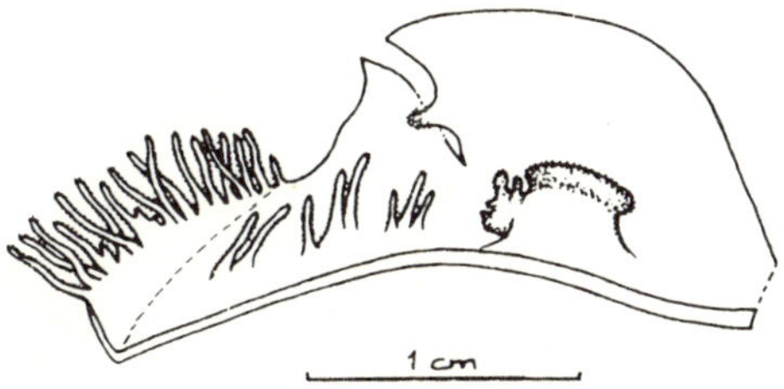

Lip, sectioned

MORMODES CONVOLUTUM LINDL.

Distribution: Venezuela and Colombia. In Venezuela it has been found as an epiphyte in the Perijá region of the state of Zulia. The only plant available for study was brought to the Tocuco Mission by a Motilone Indian. The altitude at which it was growing is thus uncertain, but was probably around 500m. The Orchid Family as a whole is characterized by having flowers with bilateral symmetry, i.e. with the left half a mirror image of the right half, but the genus *Mormodes* is an exception to this as its column has a very marked twist so that the apex lies along one side of the fleshy lip.

MORMODES CONVOLUTUM

ODONTOGLOSSUM SPECTATISSIMUM

ODONTOGLOSSUM SPECTATISSIMUM LINDL.

Synonym: *Odontoglossum triumphans* Rchb.f.
Distribution: Venezuela and Colombia. In Venezuela this large and showy species occurs as an epiphyte at 2300 to 2500m. in the state of Táchira.

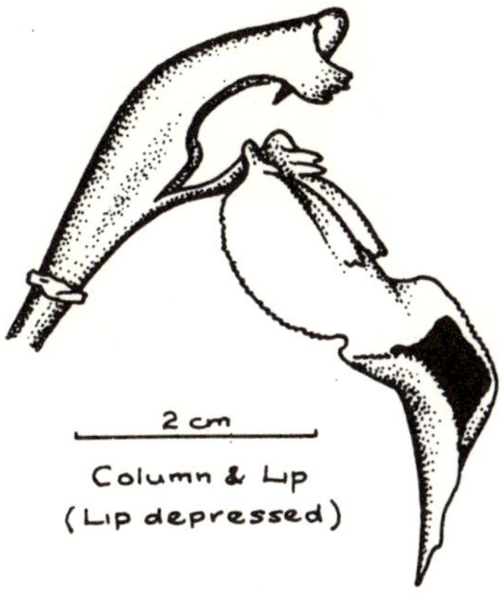

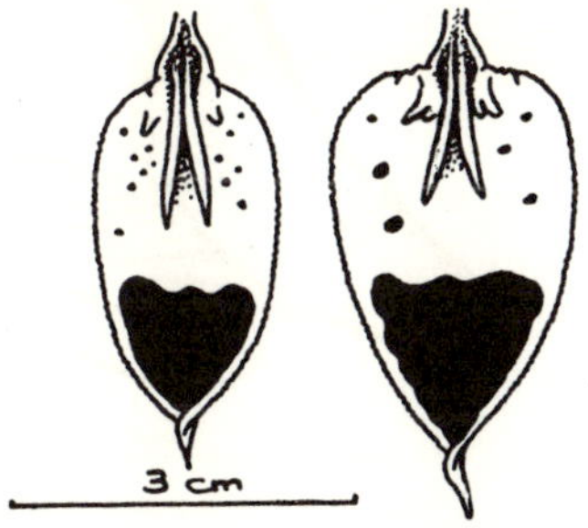

ONCIDIUM HASTILABIUM (LINDL.) GARAY & DUNSTERV.

Synonym: *Odontoglossum hastilabium* Lindl.

Distribution: Venezuela, Colombia, Ecuador, Peru. In Venezuela it has been found as an epiphyte at an altitude of 1500m in the Sierra de Perijá in the state of Zulia.

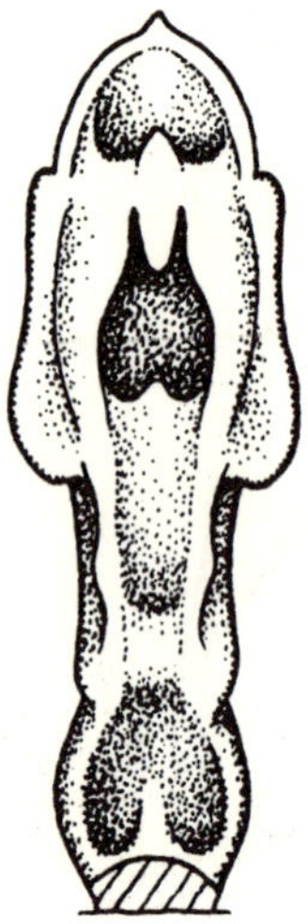

Column Face

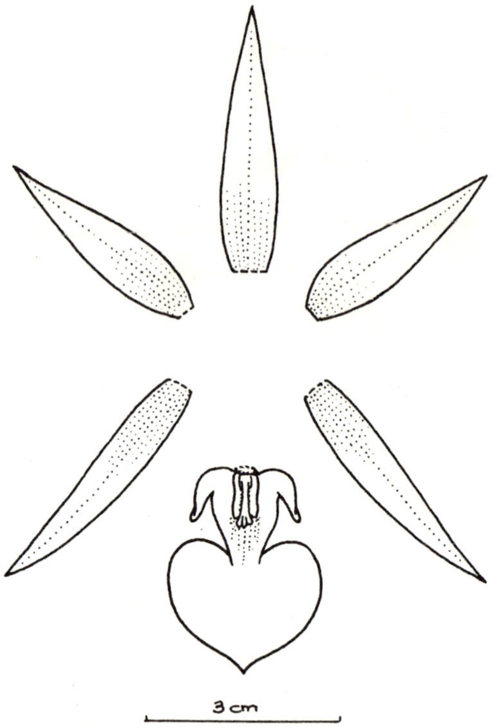

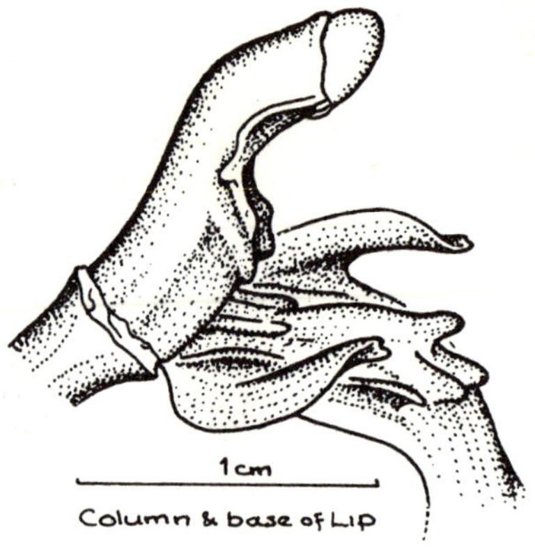

Column & base of Lip

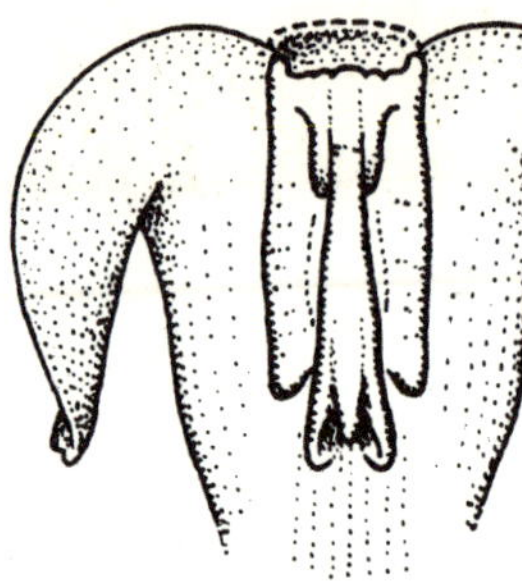

Base of Lip

ONCIDIUM HASTILABIUM

PHRAGMIPEDIUM KAIETEURUM

PHRAGMIPEDIUM KAIETEURUM (N.E.BR.) GARAY

Synonyms: *Selenipedium kaieteurum* N.E.Br.
Phragmopedilum lindleyanum var. *kaieteurum* (N.E.Br.) Rchb.f.

Distribution: Venezuela and Guyana. In Venezuela this terrestrial species appears to be rare, but may actually be quite common as it is easily confused with the much better known *Phragmipedium lindleyanum* which grows in the same area. So far it seems to have been found only in the state of Bolívar near to Cerro Venamo at 400m. altitude and in the Gran Sabana at some 600m.

In his most useful key to the 21 species in the puzzling genus of Phragmipedium in the Orchid Digest of July 1979, Garay first separates out four that, like *Phragmipedium kaieteurum* do *not* have long narrow petals with tapering apices, and then, using other factors, reduces the choice to either *Phrag. lindleyanum* with long leaves or *P. kaieteurum* with short leaves, as far as readily visible features are concerned. The drawings below show the shape of the sepals and of one of the typically blunt-ended petals, and two not readily visible details of the column.

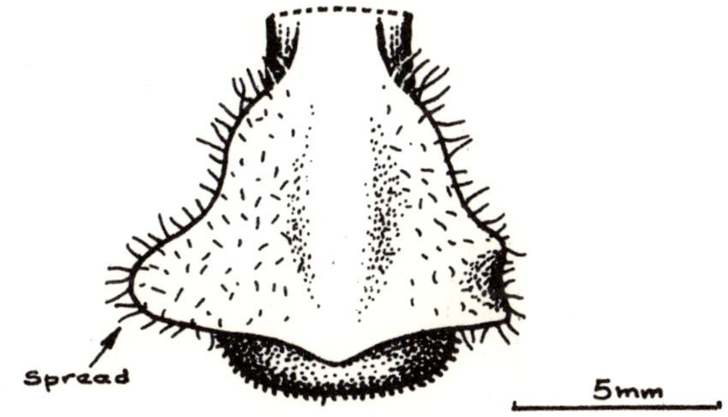

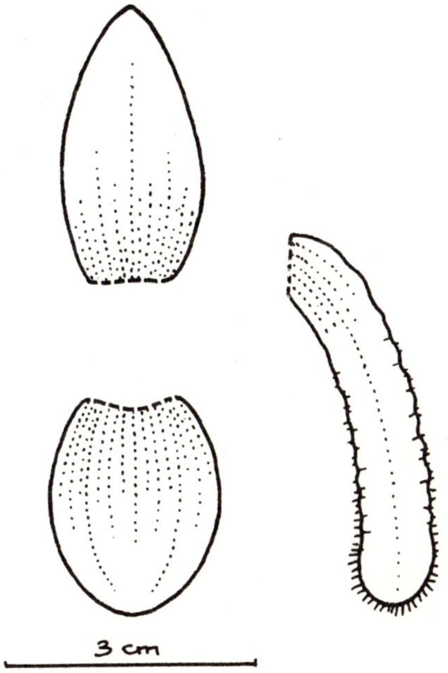

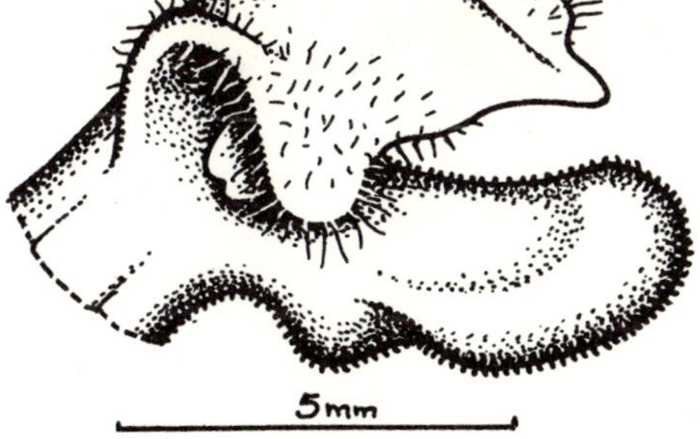

PLEUROTHALLIS DUNSTERVILLEI FOLDATS

Distribution: Venezuela, Colombia. In Venezuela this species has been found as an epiphyte below the Páramo de Tamá in the state of Táchira at an altitude of 2700m. The genus *Pleurothallis* is characterized mainly by small plants and flowers, but this species, though its flowers are small, has erect secondary stems at times nearly 50cm. tall with 15cm. leaves.

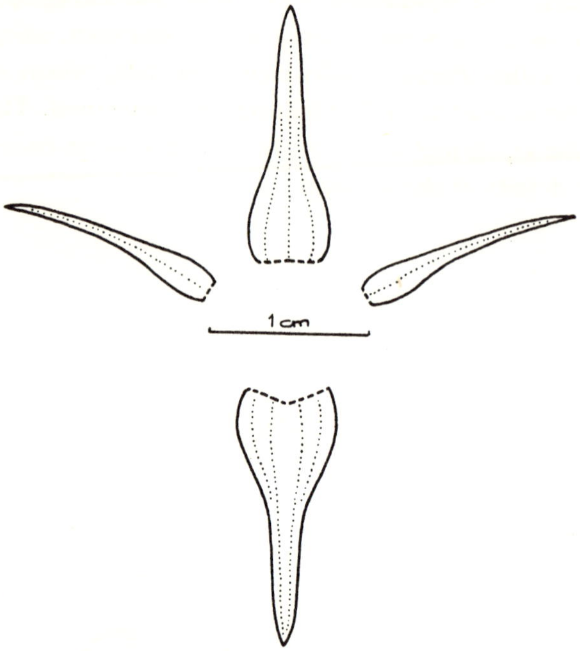

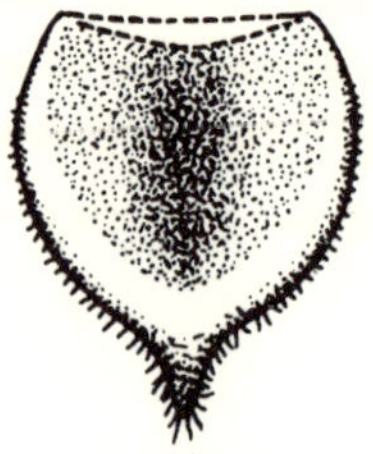
3mm

Lip

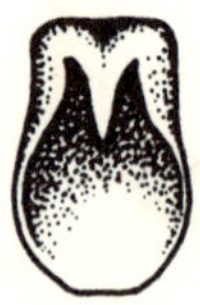
1mm

Anther

1mm

Pollinia

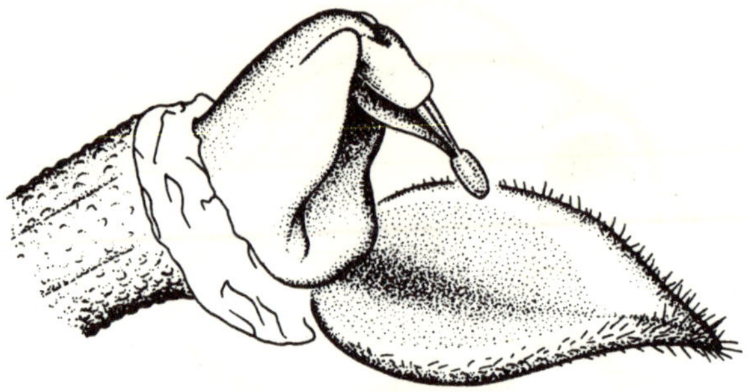
3mm

Column & Lip

PLEUROTHALLIS DUNSTERVILLEI

SOBRALIA CATTLEYA

SOBRALIA CATTLEYA RCHB. F.

Distribution: Venezuela and Colombia. In Venezuela this species can be found growing in the soil of the Andes in the state of Táchira at about 1800m. altitude. This species is notable not only for the unusual color of the flowers but for the length of the stems, which commonly exceed 2m in height and are reported to have reached more than 5m.

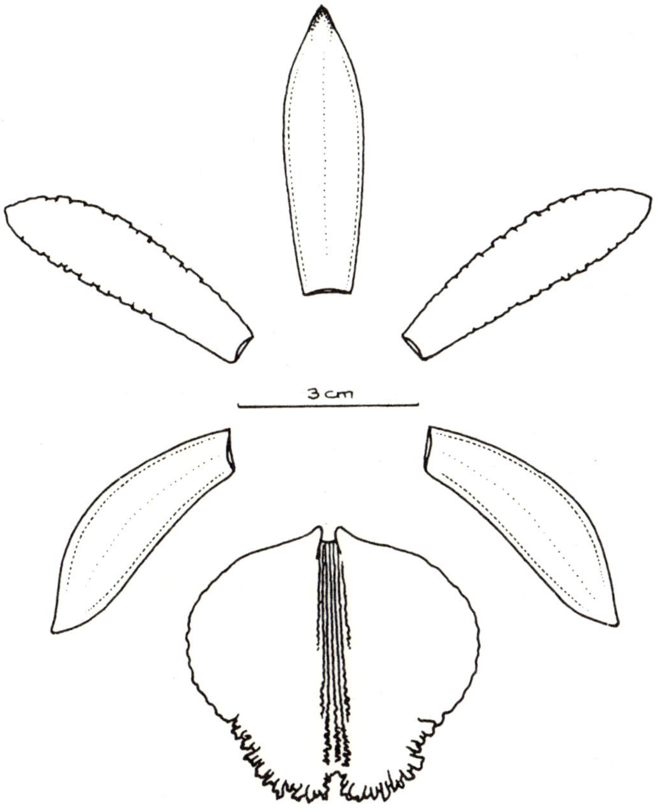

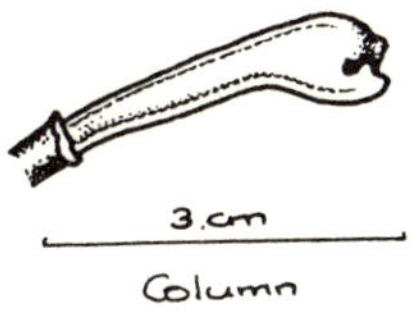

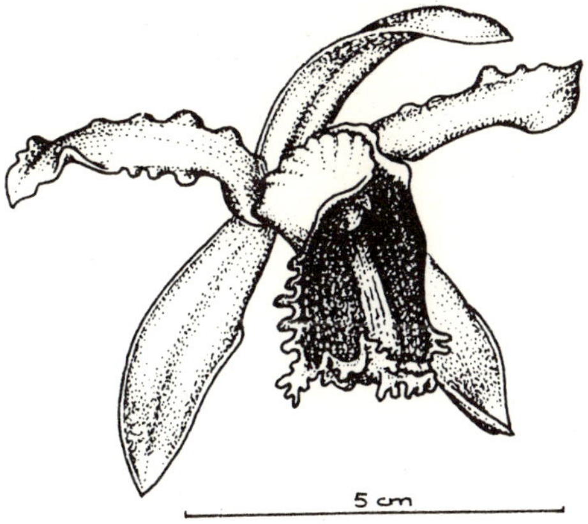

HUNTLEYA LUCIDA (ROLFE) ROLFE

Synonyms: *Zygopetalum lucidum* Rolfe
Distribution: Venezuela and Guyana. In Venezuela it is known principally from the Guatopo National Park near Caracas at about 600m. and from the Paria Península of Sucre state at about 800m. and from the Island of Margarita at around 500m. The plant takes the form of individual pseudobulbless leafy growths spaced closely along the rhizome and concealing it, but separate growths can also survive very well by themselves.

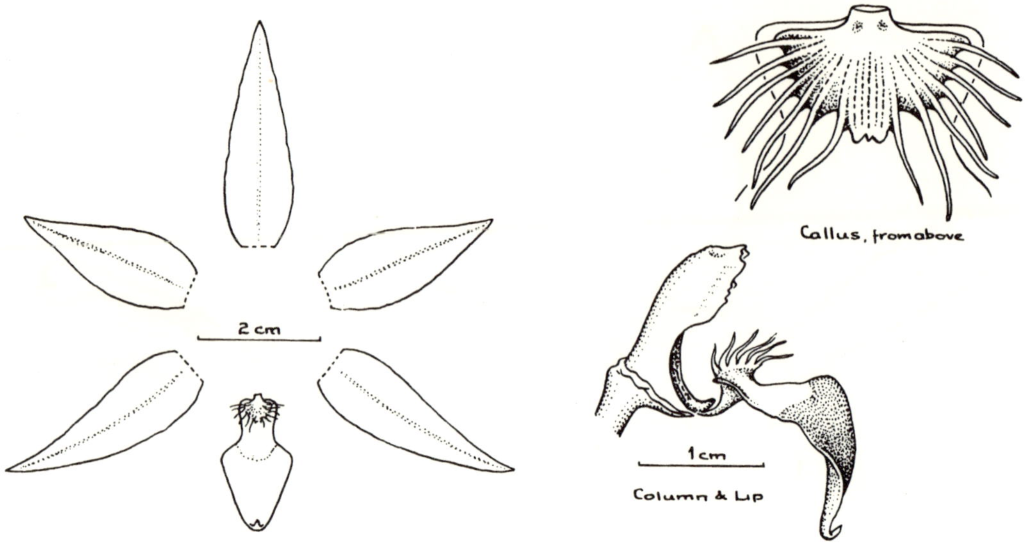

ZYGOSEPALUM LINDENIAE (ROLFE) GARAY & DUNSTERV.

Synonyms: *Zygopetalum lindeniae* Rolfe
Menadenium lindeniae Cogn.
Distribution: Venezuela and Brazil. In Venezuela this attractive epiphyte can be found at altitudes of 100 to 200m. mainly in the Upper Orinoco area from San Fernando de Atabapo to Río Padamo and beyond, including the Casiquiare. It has also been collected from near the top end of the lower Río Siapa at about 300m. altitude.

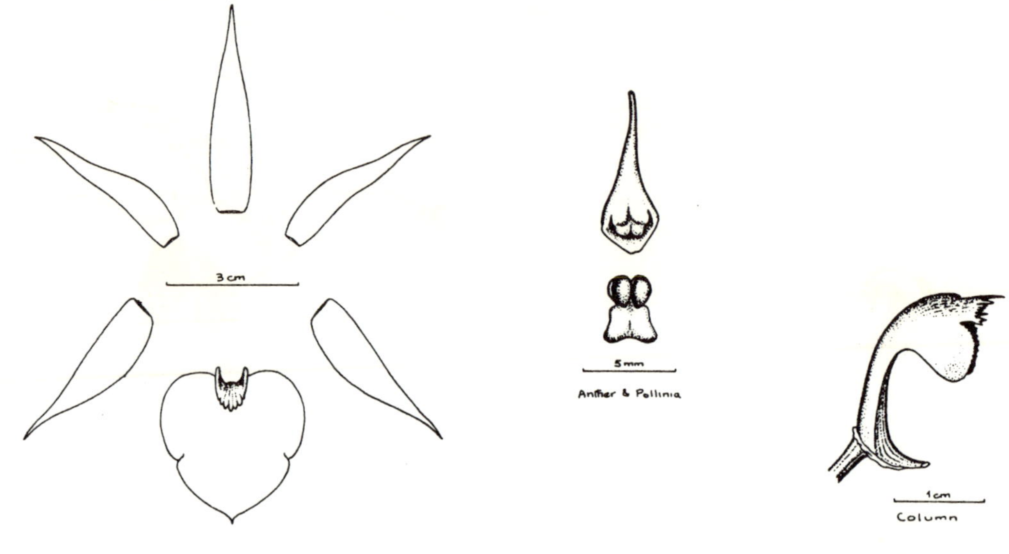

HUNTLEYA LUCIDA

ZYGOSEPALUM LINDENIAE

MAXILLARIA LAWRENCEANA

MENDONCELLA BURKEI

MAXILLARIA LAWRENCEANA (ROLFE) GARAY & DUNSTERV.

Synonym: *Camaridium lawrenceanum* Rolfe
Distribution: Endemic to Venezuela, where it has been found as an epiphyte at altitudes from 1500 to 2000m. in the states of Mérida and Táchira.

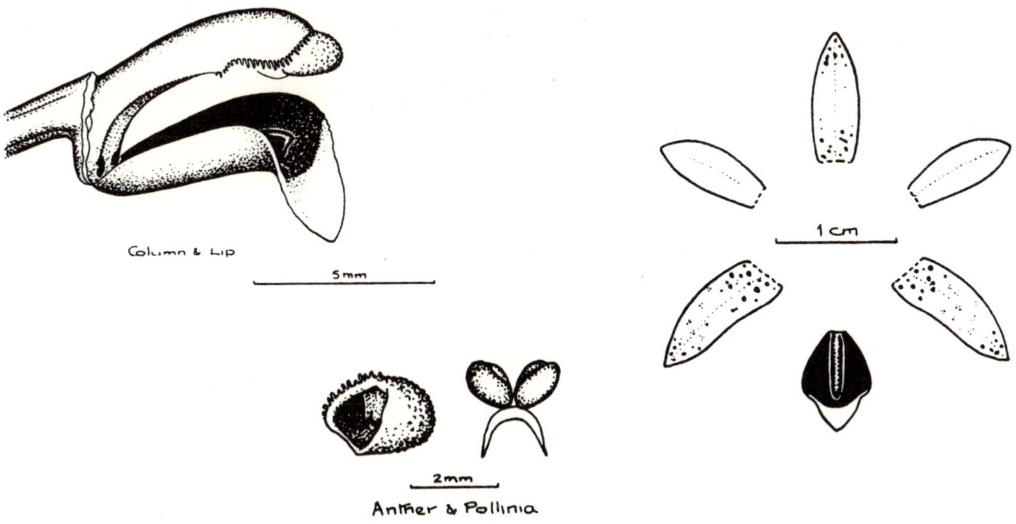

MENDONCELLA BURKEI (RCHB.F.) GARAY

Synonyms: *Zygopetalum burkei* (Rchb.f.) Garay
Zygopetalum prainianum Rolfe
Distribution: Venezuela, Guyana, Brazil, Peru. In Venezuela it is found mainly as a terrestrial orchid in Bolívar state, from 1000 to 2200m, mainly on sandstone tepuis and cerros, and in the Federal Territory of Amazonas at 1300 to 1600m on cerros such as Duida and Sipapo.

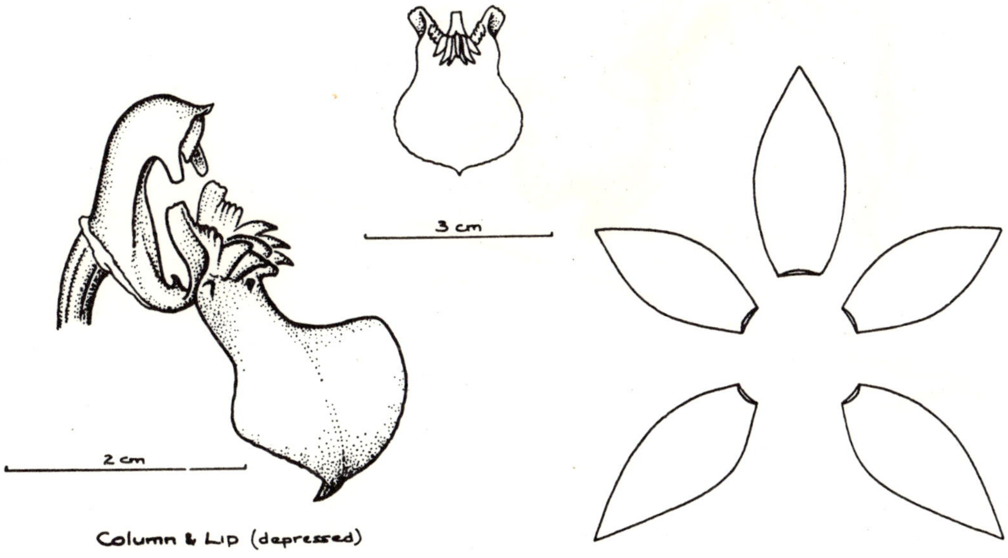

SOBRALIA INFUNDIBULIGERA GARAY & DUNSTERV.

Distribution: Venezuela, endemic. This terrestrial species has been found in the state of Bolívar in a number of localities such as at 400m. on the Río Aonda to 2000m. on the summit of Auyantepui, with intermediate altitude records from Cerro Guaiquinima, Cerro Jaua and Cerro Venamo. It also occurs in the Federal Territory of Amazonas at 1400m. on Cerro Duida.

The plants have robust stems and leaves with wide, sharp-edged and funnel-shaped bases that sheath the stem and give it its name of "infundibuligera". Besides the colored form shown here, this species also produces on occasion a pure white form.

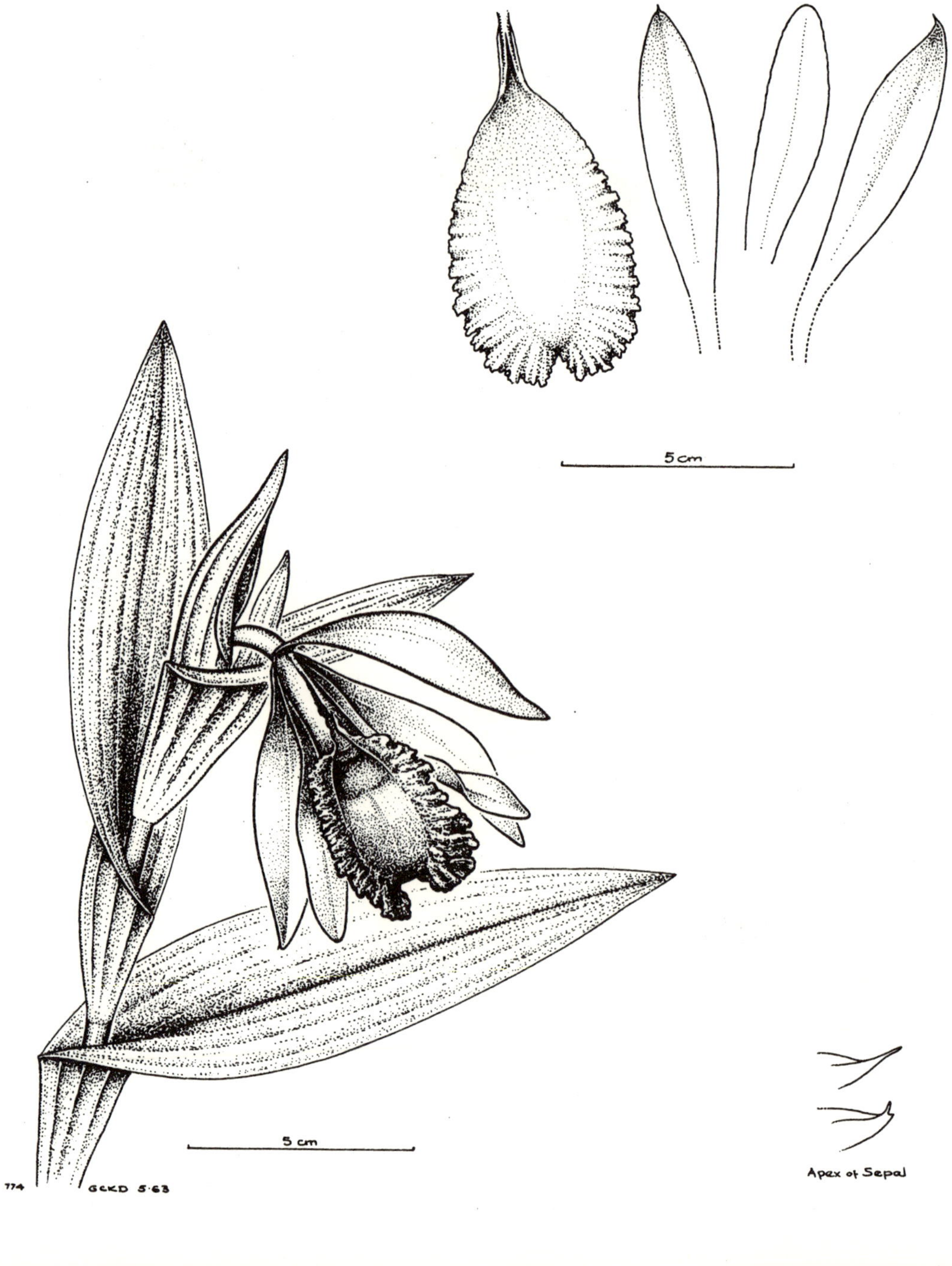

SOBRALIA INFUNDIBULIGERA

SOBRALIA RUCKERI

SOBRALIA RUCKERI LINDEN & RCHB.F.

Distribution: Venezuela and Colombia. In Venezuela this species has been found growing as a very robust terrestrial at about 2000m. altitude in the Sierra de Perijá in the state of Zulia.

This species is notable not only for the beauty of its flowers but also for their long-lasting character, and for the fact that up to six flowers can be borne on a single raceme — a pleasant contrast with most Sobralia flowers which are developed singly and seldom last more than a day.

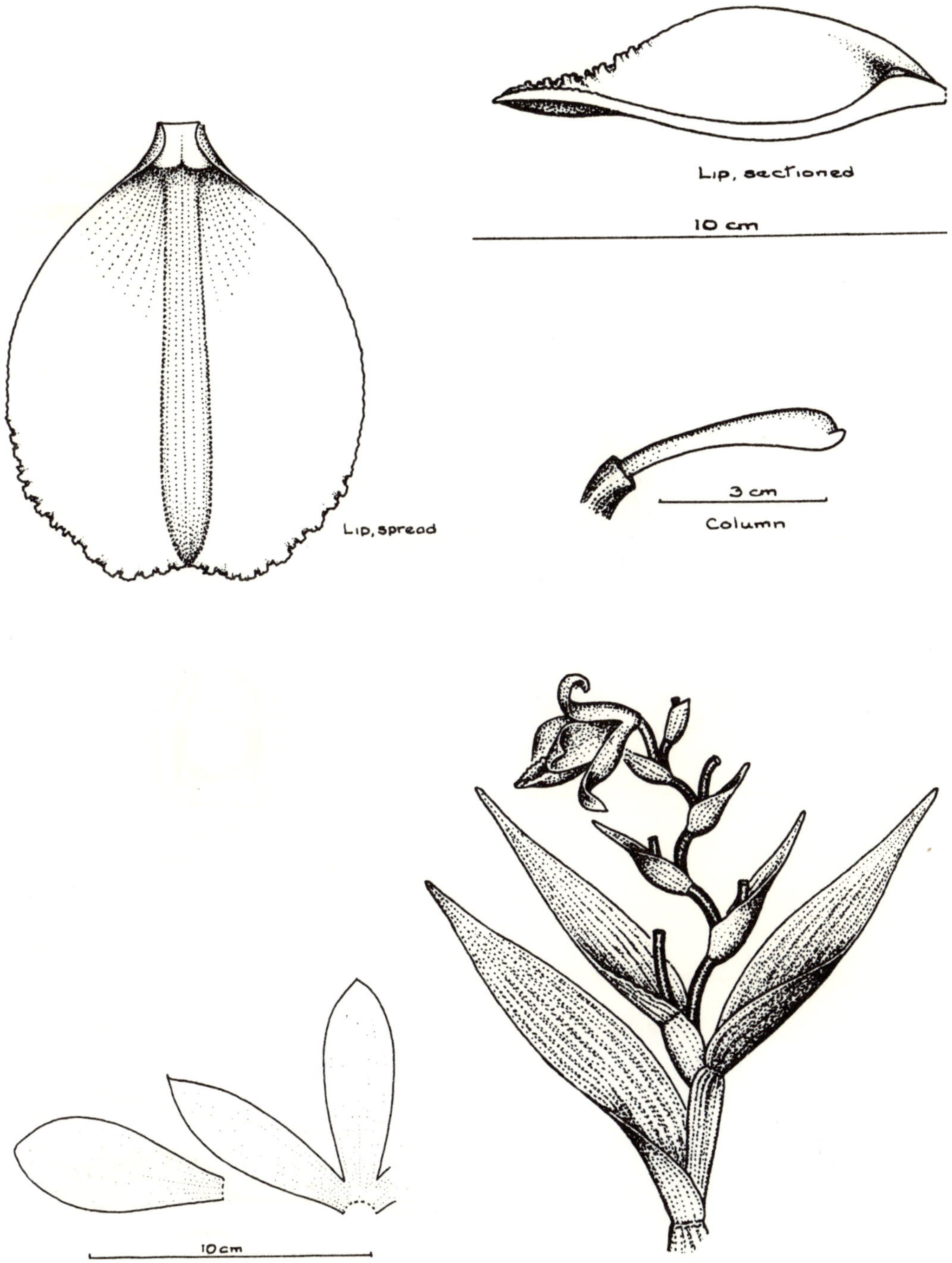

SOBRALIA VIOLACEA LINDEN

Synonyms: *Cattleya odoratissima* P.N. Don
Cattleya violacea Beer

Distribution: Venezuela, Colombia, Peru, Bolivia. In Venezuela it is a common terrestrial orchid in the Andes at altitudes of 1000m to 2400m, and is also present at 300m to 1300m altitude in the Perijá region of the state of Zulia. In the state of Bolívar it has been found at as low as 450m on the Río Aonda, and in the Federal Territory of Amazonas a very pale form was found on the Río Siapa at about 500m. The apex of the stem, just below the uppermost leaves, is characterized by rough, almost warty, sheathing leaf bases.

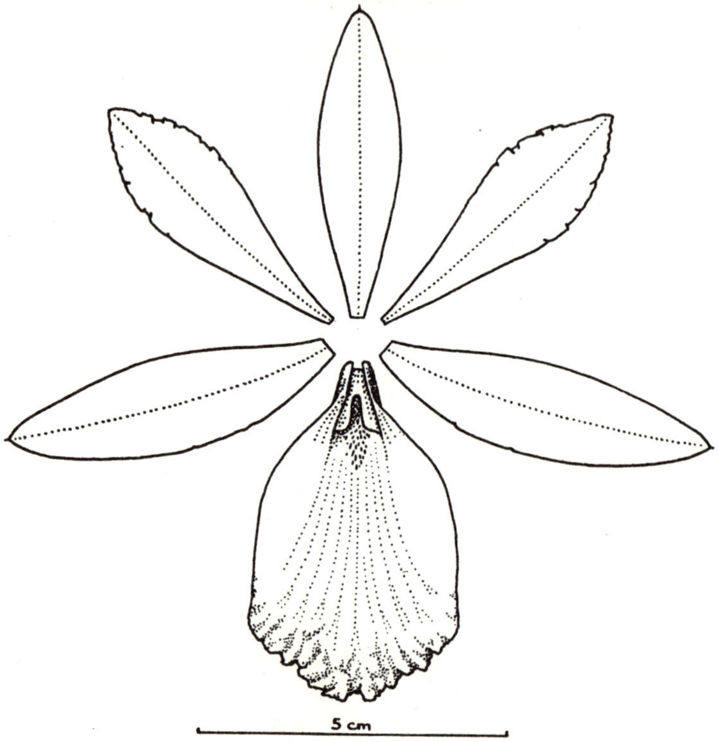

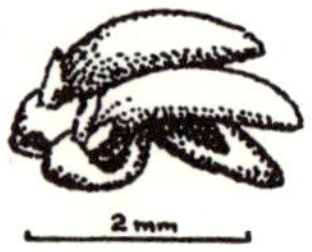

Pollinia

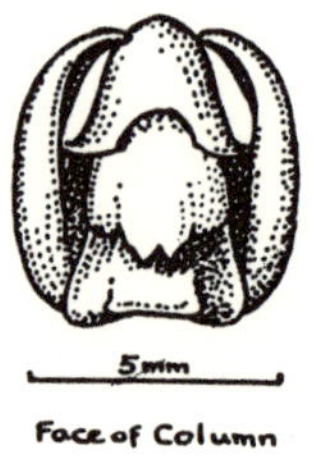

Face of Column

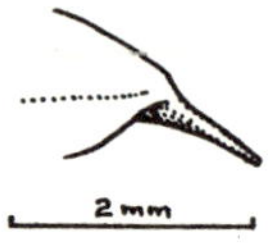

Apex of Sepal

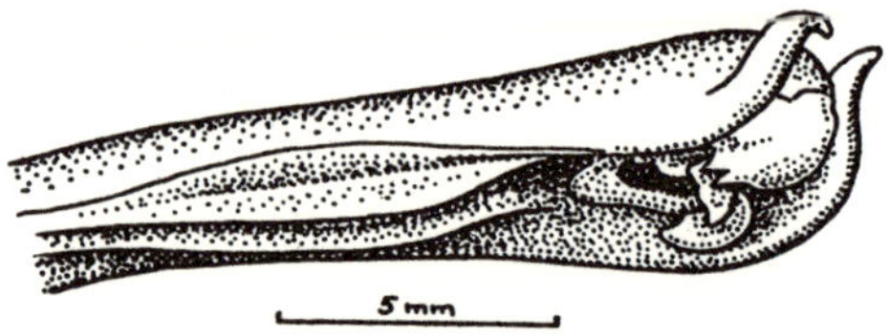

Underside of Column

SOBRALIA VIOLACEA LINDEN

STANHOPEA GRANDIFLORA

STANHOPEA GRANDIFLORA (LODD.) LINDL.

Synonyms: *Stanhopea eburnea* Lindl.
Ceratochilus grandifloris Lodd.

Distribution: Venezuela, Trinidad, Guyana, Colombia, Brazil. In Venezuela this epiphytic species can be found in the Guatopo National Park of the state of Miranda at about 800m. altitude, and at about 1000m. altitude in the Paria Península in the state of Sucre. In the state of Bolívar it occurs at altitudes of 400 to 600m. on the edge of the Gran Sabana, and in the Federal Territory of Amazonas it can be found at altitudes of 120 to 300m.

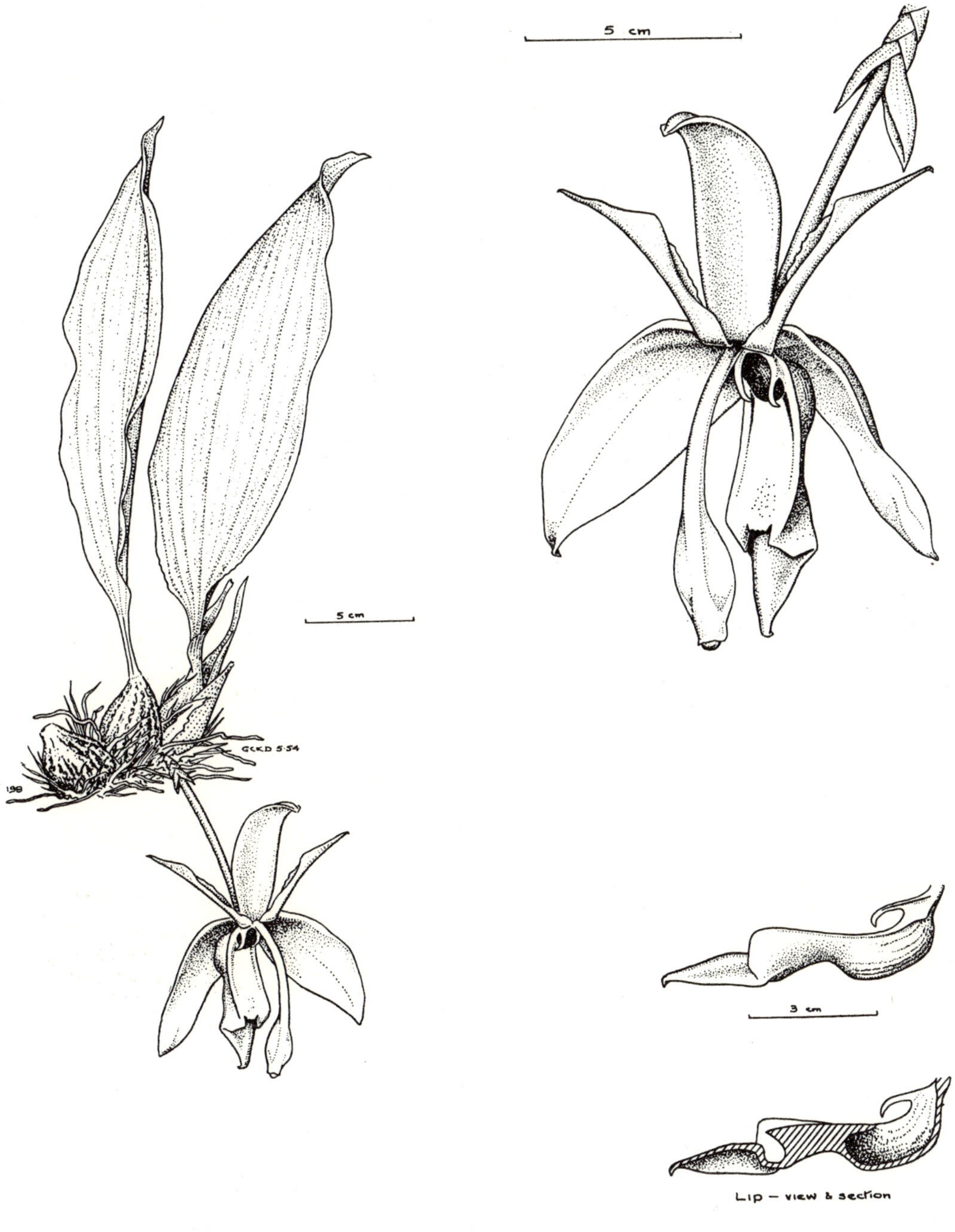

Graphic design
Michele Ferrelli

Drawings and photography
Galfrid C.K. Dunsterville

Spanish translation
Frances McCarty

Color separations and printing
Gráficas Armitano, C.A.

English edition
1,000 Copies

Spanish edition
1,500 copies

Paper
Plates Ikonofix 100 lbs.
Drawings Hamilton 70 lbs.

Types
Garamond 12 y 14 points
Roman, italic and bold

August 28, 1987

Gráficas Armitano, C.A.

Tel: 34.25.65 master - 34.08.65 - 34.08.70